THE APPLICATION OF MATHEMATICS IN INDUSTRY

THE APPLICATION OF MATHEMATICS IN INDUSTRY

Edited by

Robert S. Anderssen
Division of Mathematics and Statistics,
CSIRO, Canberra City, ACT 2601

and

Frank R. de Hoog
Division of Mathematics and Statistics,
CSIRO, Canberra City, ACT 2601

1982

SPRINGER-SCIENCE+BUSINESS MEDIA, B.V.

Library of Congress Cataloging in Publication Data
Main entry under title:

The Application of mathematics in industry.

   "Proceedings of a one-day seminar on the
application of mathematics in industry held at
the Australian National University ... organized
jointly by the Division of Mathematics and Statistics, CSIRO, and the Departments of Pure and
Applied Mathematics, the Faculty of Science,
Australian National University"--Foreword.
   1. Engineering mathematics--Congresses.
2. Manufacturing processes--Mathematical models--
Congresses. I. Anderssen, R. S. II. De Hoog,

Frank R. III. Commonwealth Scientific and Indus-
trial Research Organization (Australia). Division
of Mathematics and Statistics.
TA329.A66   620'.0042   82-2149
ISBN 978-94-011-7836-5   ISBN 978-94-011-7834-1 (eBook)
DOI 10.1007/978-94-011-7834-1

TABLE OF CONTENTS

FOREWORD

This publication reports the proceedings of a one-day seminar on *The Application of Mathematics in Industry* held at the Australian National University on Wednesday, December 3, 1980. It was organized jointly by the Division of Mathematics and Statistics, CSIRO, and the Departments of Pure and Applied Mathematics, The Faculty of Science, Australian National University. A paper based on the talk "Some uses of statistically designed experiments in industrial problems" given by N.B. Carter at the Seminar was not received by the editors. Though R.M. Lewis of John Lysaght (Australia) Limited did not present a talk, the editors invited him to submit a paper. They only learnt about his work after the program for the seminar had been finalized and publicized. His paper appears as the last paper in these proceedings and is entitled "A simple model for coil interior temperature prediction during batch annealing".

The seminar was opened by Dr J.R. Philip, FAA, FRS, Director of the Physical Sciences Institute, CSIRO. He kindly agreed to supply an edited version of his comments for inclusion in the proceedings. They follow the Foreword as Opening Remarks.

An attempt was made to structure the program to the extent that (a) a major aim in organizing the seminar was to bring together the academic mathematician with little first hand experience with the application of mathematics to industrial problems, and non-academics with an ongoing industrial responsibility; (b) the seminar itself aimed to illustrate the different reasons why mathematics should be, is or must be used in the solution of industrial problems; and (c) all speakers were requested to organize their talks so that they first discussed an application, then showed why mathematics was necessary for the solution process, and only then discussed the mathematics itself.

The organizers would like to take this opportunity to record their appreciation and thanks for the contributions of the speakers, Dr John Philip for his Opening Remarks, the able assistance of the session chairmen (Dr Chris Heyde, Division of Mathematics and Statistics, CSIRO; Professor Archie Brown, Department of Applied Mathematics, The Faculty of Science, Australian National University; Professor Neil Trudinger, Department of Pure Mathematics, The Faculty of Science, Australian National University; and Dr John Philip, Director, Institute of Physical Sciences, CSIRO) and the active involvement of the participants. Thanks must also go to the Division of Mathematics and Statistics, CSIRO, and the Departments of Pure and Applied Mathematics, The Faculty of Science, Australian National University, for their support and assistance without which the Seminar could not have been the success that it was.

The excellent work of Barbara Harley and Dorothy Nash in assisting with organizational aspects associated with both the Seminar and Proceedings, of Clementine Krayshek who traced many of the figures and of Anna Zalucki who typed the proceedings is gratefully acknowledged with sincere thanks.

R.S. Anderssen and F.R. de Hoog
Canberra, ACT 2600
June, 1981

OPENING REMARKS

Let me begin by applauding the CSIRO Division of Mathematics and Statistics, and the mathematicians of the A.N.U., for their initiative in bringing into being this One-Day Seminar on Mathematics in Industry. So far as I can discover, the insight and the work which have led to this occasion have been provided largely by Bob Anderssen and Frank de Hoog. We are all most grateful to these two, and to everyone else involved in the preparations for the Seminar.

I have been commanded to provide some Opening Remarks. I am at something of a loss, since there is already in existence the Introduction which follows these remarks. That Introduction seems to say very well most things that might conceivably be said in Opening Remarks. So I shall do no more here than offer you one or two brief observations.

To begin with, I believe that one cannot repeat often enough just what is involved in applying mathematics to real world problems, whether in industry or in the wider context of natural science. Firstly, we must make a proper abstraction based on the real world problem. We must have sufficient insight into what is happening, and sufficient empirical knowledge of it, to be able to sort out the significant elements in the problem. We must not neglect important factors; and yet, on the other hand, we must not retain extraneous elements which make things complicated, obscure and difficult, but shed no light on the major issues.

Secondly, we must be able to set up the abstract problem so that it is amenable to the processes of mathematics.

And thirdly, we must be able to translate our solutions of the abstract problem back to the real world accurately, and with a clear and correct recognition of the limitations to our solution.

The Irish mathematician, J.L. Synge, described this three-stage process vividly and succinctly as follows :

"A dive from the world of reality into the world of mathematics; a swim in the world of mathematics; a climb from the world of mathematics into the world of reality, carrying a prediction in our teeth."

May I say that I find the enthusiasm for the application of mathematics in industry manifest in this seminar most commendable. I should like, however, to say that, in my opinion, we can do with considerably more enthusiasm directed towards the use of mathematics in the related but separate field of natural science in general.

It is my personal view that a great number of natural scientists abandon their mathematics far too early in their training, and live out their careers deprived of a central tool of their trade. I believe that one reason for this is the way in which mathematics is often taught.

As the following Introduction hints, many good natural scientists find their motivation, not in the immediate pursuit of maximum generality, but in the process of working from the particular to the general.   As many of you know, that viewpoint arouses only the disdain of many of our professional mathematicians.   These, unfortunately, tend to be the people who teach mathematics to our potential natural scientists.

Let me close with a quotation from Professor Kaplan which, I am sure, has meaning for all of us; but I expect it holds special significance for the modellers amongst us.

"Models are undeniably beautiful, and a man may justly be proud to be seen in their company.   But they may have their hidden vices.   The question is, after all, not only whether they are good to look at, but whether we can live happily with them"!

J.R. Philip

INTRODUCTION

"To explain the methods that have been devised to bring mathematics into some aspects of the everyday world is a more difficult task than to explain mathematics itself".

Norman Clark,
Secretary and Registrar, IMA,
in Foreword of
J. Lighthill, Newer Uses of Mathematics.

In examining "the application of mathematics in industry", it is not the range of mathematical techniques used which must be discussed, but the nature of how mathematics interacts with applications to the benefit of both. As the above quotation implies, that is easier said than done. The difficulty is the unstructured diversity of industrial applications that are amenable to mathematical analysis.

In order to give meaning to "The Application of Mathematics in Industry", it is necessary to choose from the huge range of possibilities a subset which reflects the utility of mathematics in industry.

In this seminar, we have chosen applications which illustrate the different reasons why mathematics should be, is or must be used. An intuitive understanding of the application of mathematics in industry is based on an appreciation of such reasons, and not solely on the mathematical tools applied. These include :

(i) For the optimal utilization of large and complex plant which cannot be modelled experimentally (such as a blast furnace), mathematical modelling is the only alternative.

(ii) In the examination of properties of new materials (such as composites), the use of mathematical models often avoids the necessity to perform long and/or expensive experiments.

(iii) The use of statistically designed experiments can often ensure that the control and monitoring of industrial processes is done with maximum efficiency.

(iv) Through the use of computational techniques, an increasing level of mathematical sophistication has been brought to bear on industrial processes.

(v) Industrial problems which can be simply explained do not necessarily have equally simple answers and can involve the use of highly sophisticated mathematics.

(vi) The efficiency and effectiveness of non-destructive testing and analysis.

(vii) The fact that, in many applications, the information required only comes from indirect measurement.

(viii)   Even though the mathematics involved may be quite elementary, its need may be
         crucial in ensuring that an industrial process operates correctly.

Globally, applied mathematics and the application of mathematics in industry are similar
in that they involve the three basic but interrelated steps of : (i) formulation; (ii) solution;
and (iii) interpretations.   However, as with the different branches of science, the inference
patterns used in the application of mathematics to industrial problems often differ from those
used in other branches of mathematics.   These differences depend heavily on factors like :

(a)   Compared with applied mathematics, which aims, through the use of mathematics, to
      seek knowledge and understanding of scientific facts and real world phenomena, the
      application of mathematics in industry aims to answer specific questions of immediate
      and direct concern.

(b)   Compared with the academic situation where one has "a method without a problem", the
      situation in industry is very much one where one has "a problem without a method".
      In part, this is why computers have had such a big impact on applications as they
      allow, in no other way possible, for the problem to remain of central importance.

(c)   As long as the given problem is solved economically and effectively, the lack of
      mathematical sophistication or generality is of no major concern.   The aim is to
      choose a framework in which one can rigorously answer the relevant questions.

(d)   One of the major difficulties in solving industrial problems is ensuring right from
      the start that the question being answered is the question which should have been
      asked.   The form of the mathematics used depends heavily on the question which
      must be answered.

Overriding all these points is the fact, which does not appear to be fully appreciated,
that much useful mathematics is simple in nature.   As a consequence, the average mathematician
can contribute to the application of mathematics, as long as he has the motivation and
interest.   This point has been made indirectly by Jeffreys and Jeffreys in the Preface to
their book on "Methods of Mathematical Physics" :

> "We think that many students ... have difficulty in following abstract arguments,
> not on account of incapacity, but because they need to "see the point" before
> their interest can be aroused."

SUMMARY OF THE TALKS

The Director, Physical Sciences Institute, CSIRO, Dr J.R. Philip, FAA, FRS, launched the
Seminar with some opening remarks about the role of mathematics in industry.   In the first
talk, Dr John Lowke of the Division of Applied Physics, CSIRO, highlighted the fact that, in
industrial problems where many physical processes occur simultaneously, it is difficult to
identify which process, if any, dominates. Dr Lowke used two industrial examples to show how
detailed mathematical modelling and experimentation can be used to identify the processes
which dominate.   In the first, the problem of determining the operating voltages of lasers was

discussed; while, in the second, the prediction of the diameters of arcs which occur in circuit breakers was examined.

The next speaker was Dr Rys Jones of the Aeronautic Research Laboratories, Melbourne. His talk, which was about the mathematical analysis of *in situ* repairs of cracked aircraft components, illustrated clearly the need to use mathematical tools to avoid the need to perform long and/or expensive experiments. The next three talks by Mr David Jenkins, Dr Peter Swannell and Dr Val Pinczewski were all concerned in various ways with the use of computational tools in order to increase the level of mathematical sophistication that is brought to bear on industrial processes. Mr Jenkins spoke about the mathematical modelling of gas flow in blast furnaces; Dr Swannell discussed the dynamic behaviour of the Gateway Bridge just approved for construction across the Brisbane River; and Dr Pinczewski examined the simulation of lateral liquid flow in the hearth of a blast furnace. In different ways, they also illustrated points already made by the first two speakers as well as the fact that, for the optimal utilisation of large and complex plant, mathematical modelling is the only alternative.

The talk by Dr Ray Volker about a mathematical model for the storm water drainage system in Townsville, Queensland, aimed, among other things, at illustrating that practical problems which can be simply explained do not necessarily have equally simple answers, since the complexity of the underlying practical problem is not easily modelled mathematically.

Dr Greg Taylor then discussed why the survival of Non-Life Insurance Companies depends heavily on the sophistication and accuracy of their actuarial modelling. Dr Bob Johnston examined the mathematical methods which were now being used widely throughout industry to maximise the efficiency of the cutting of stock. Ms Nan Carter and Dr George Brown then described how statistically designed experiments can often ensure that the control and monitoring of industrial processing is done with maximum efficiency. Ms Carter explained the use of such work in the monitoring of ore separation machines, while Dr Brown discussed the use of acceptance sampling for the detection of bad batches with an example from the peanut industry.

In his talk about the grinding of contact lenses, Dr Bill Davis aimed at illustrating the important point that, though the mathematics involved may sometimes be elementary, the need for it to be correct is crucial for the success of the underlying industrial process. A further illustration of the point that, though industrial problems can often be simply explained, the underlying mathematics may be highly sophisticated was given by Dr Richard Cowan in his talk about sheet metal bending. The Seminar finished with Dr Bob Anderssen's talk about the fabrication of optical fibres. It was primarily concerned with the fact that, in many applications, the required information only comes from indirect measurement.

BACKGROUND READING

For people also interested in the application of mathematics in a wider context than covered by these proceedings, the following non-exhaustive list of references can be used as starting points for many of the possible directions that can be pursued.

J.G. Andrews and R.R. McLone (Editors), *Mathematical Modelling*, Butterworth, London, 1976.

E.A. Bender, *An Introduction to Mathematical Modelling*, J. Wiley and Sons, New York, 1978.

J. Crank, *Mathematics and Industry*, Oxford University Press, London, 1962.

J. Lighthill (Editor), *Newer Uses of Mathematics*, Penguin Books, Harmondworth, Middlesex, England, 1978.

C.C. Lin and L.A. Segal, *Mathematics Applied to Determmistic Problems in the Natural Sciences*, Macmillan, New York, 1974.

C.C. Lin, On the role of applied mathematics, *Advances in Math* 19 (1976), 267-288.

C.C. Lin, Education of applied mathematicians, *SIAM Review* 20 (1978), 838-845.

F.J. Murray, *Applied Mathematics : An Intellectual Orientation*, Plenum Press, New York and London, 1978.

J.R. Ockendon, Differential equations in industry, *The Math. Scientist* 5 (1980), 1-12.

D.J. Prager and G.S. Omenn, Research, innovation and university-industry linkages, *Science* 207 (1980), 379-384.

P.C.C. Wang, A.L. Schoenstadt, B.I. Russak and C. Comstock (Editors), *Information Linkage Between Applied Mathematics and Industry*, Academic Press, New York, 1979.

R.S. Anderssen and F.R. de Hoog
Canberra, ACT 2600
June, 1981

PREDICTION OF OPERATION VOLTAGES
OF CO_2 LASERS AND LIMITING CURRENTS
OF CIRCUIT BREAKERS

J.J. Lowke,
Division of Applied Physics, CSIRO, Sydney.

ABSTRACT

A major difficulty in doing calculations for industrial problems is that usually many physical processes occur simultaneously. However a detailed analysis frequently reveals that one process is dominant and that, as a consequence, approximate mathematical expressions can be derived that are adequate for the engineer. Two examples are given. Firstly, for CO_2-N_2-He gas lasers a two-term expansion of the electron distribution function in spherical harmonics yields electron attachment and ionization coefficients from the Boltzmann transport equation. Equating ionization and attachment at equilibrium enables predictions of the operating voltage of the laser to be made. Secondly, detailed solutions of the radiative transfer equation make it possible to predict the diameters of arcs that occur in circuit breakers in a simple way.

1. INTRODUCTION

There are a large number of differences between the university environment where a mathematician or physicist obtains his training, and the industrial environment where he must work and make a living. These differences I found to be rather daunting, even when moving to the relatively sophisticated environment of an American industrial research laboratory. For a mathematician working in Australian industry, the differences would be harsher.

It is of value to delineate some of the differences between university and industrial environments :

(i)   In a university, courses in mathematics, physics and engineering tend to be dominated by theory.   Ones colleagues at university understand, or at least can potentially understand, what one is doing.   Furthermore, the output of the university is largely in the form of papers, where theory usually plays a dominant role.

On the other hand, industry is dominated by hardware, concrete and devices.   The function of industry is not to produce theory or papers, but to sell devices or products and make a profit.   In industry the mathematician, and physicist are in a minority.   The mathematician's manager and most of his colleagues will have little appreciation for the sophisticated mathematics with which he is working.   The mathematician or physicist is often regarded as being a parasite on the engineering establishment and the onus is on him to justify his existence.

(ii)   In a university, the undergraduate or PhD student is given a reasonably defined problem. Furthermore the undergraduate knows that there are answers to these problems.   A PhD student, provided his PhD supervisor is competent, knows that by using reasonably established methods, a given field can be extended.   Even if a given approach is fruitless, a research paper can result and he is credited with a degree of success for professionally competent work.

In industry, one is presented not with a simple defined problem, but dozens of problems, all of which appear to be completely intractable and to have no rational solution.   The engineers designing gas lasers want to get answers to the following types of questions : - what voltage power supply is needed to drive a laser of given dimensions, how are these requirements varied for different mixtures of gases, would the laser efficiency be increased by varying the mixture, what is the voltage required for initial breakdown, what is the effect of water vapour impurity? etc.   The circuit breaker engineer would like to know what is the arc diameter as a function of current, would the current interrupting capability be increased by a longer or narrower nozzle, what is the effect of metal impurities from the vaporized electrodes?   The problems go on and on, each problem appearing at least at first sight, to be quite unrelated to the skills of being able to invert matrices, manipulate Laplace transforms or understand group theory.

(iii) University type problems which, for example, involve the solution of differential equations, are usually linear and involve reasonably simple geometries.   Fairly accurate solutions are possible.

Industrial problems are usually non-linear and geometries are usually complex.   Usually material functions are required which are either unknown, or, at best, are known to 20%.   Very accurate solutions are usually of no consequence - either the circuit breaker interrupts a given current or it doesn't.

In this paper, two topics are described in which mathematics was applied to solve a problem of direct industrial interest.   The first is in the prediction of the power supply requirements to operate CO_2 gas lasers used for cutting and welding, and possibly for laser fusion and military defense.   The second arises in the design of high-power circuit breakers used to protect generators in power stations.   The experiences relate to a period when the author spent 11 years working at Westinghouse in the U.S.   High technology research in secondary industry in Australia is notoriously weak, but as Australia's industry stengthens, it is expected that these U.S. experiences will become more relevant to Australian conditions.

In Section 2 we discuss the question of selecting a problem, in Section 3 the formulation of the problem in mathematical form, in Section 4 the mathematical solution, and Section 5 discusses the interpretation of the solution.

2. SELECTING A PROBLEM

The most pressing problems of any industry are usually insoluble mathematically. The laser industry may like to know the gases or combination of gases to make the most powerful laser possible. Predictions of laser performance while feasible, are dependent on a host of cross-sections for electronic excitation and atomic and molecular collisions, which just aren't known. The circuit breaker engineer would like to know whether a circuit breaker of given nozzle design in a particular circuit will interrupt a particular current. While scientific capability is tantalizingly close to being able to make such predictions, a host of uncertainties has so far denied such success.

But there are many problems in any given industry producing a particular product. The mathematician or physicist doesn't need to solve all of them, or even the most important problem, to justify his salary. In fact, because the industrial impact of success in any area is so great, he can be employed for many years just on the prospect that one day he might solve such a problem!

Usually the mathematician or physicist has some freedom to select a problem and, of course, he picks any problem that it is possible for him to solve. To be successful, he has to fulfill a double job. Firstly, he has to keep up with his subject from an academic point of view. He needs to be aware of recent developments in his subject, so that if he is a mathematician solving coupled differential equations he needs to be aware of new methods of inverting matrices, finite element methods, new computer routines that are available etc. If his problems involve electrical discharges, he needs to be aware of new insights obtained by physicists working in this field. He will need to attend specialist conferences and study current literature of the area.

Secondly, he will need to be aware of the practical problems of the engineers working in the industry. No academic can contribute to the practical problems of an industry unless he knows what these problems are. To obtain detailed knowledge of these problems is probably the most difficult task of an academic in a university or other research institution.

Furthermore, the obvious means of obtaining detailed knowledge of problems are frequently not productive. In a visit by an academic to an industry, the proud guide often does not reveal problem areas initially. Subsequent visits are likely to be regarded by the industry as a waste of their time.

Probably the best method of identifying problem areas is to go to specialist conferences There are specialist laser meetings held annually; circuit breaker engineers meet at the "Current Zero Club" an international forum that is held biennially.   For every professional group, be they welding engineers, illumination engineers or engineers interested in fuses, there is a regular society or forum where problems are discussed.   All the above areas relate to ionized gases and gas discharges and so involve plasma physics.   However even attending such conferences can be of limited value in that the papers presented merely outline solutions of problems that have been solved.   It is only in the question period or discussions over coffee that one becomes acquainted with current problem areas.

3. MATHEMATICAL FORMULATION OF PROBLEM

1. INTRODUCTION

The problems discussed in this paper involve quantitative predictions i.e. predictions of the operating voltage of a gas laser and the diameter of an arc for a given current in a circuit breaker.   Thus mathematics is an essential component of the solution process.   We assume that the phenomena that we desire to predict obey the laws of physics, and that these laws are expressible in terms of mathematical relations - usually differential equations.

The laws of physics are reasonably well established.   The behaviour of any industrial device may be the result of many complex phenomena, but it is unlikely that any new phenomena or physical processes will thwart our powers of prediction.   The problem is simply that there are too many phenomena, all occurring together, in 3 dimensions and varying in time.   Even with computers we need to make approximations and only treat the dominant physical processes.

We now discuss, in turn, the hardware and the dominant equations for these two problems.

2. HARDWARE

(1) CO_2 lasers -

The physical configuration of the electrodes of a CO_2 gas laser is shown in Figure 1. The laser is operated in the pulse mode, with a mixture of carbon dioxide, nitrogen and helium passed between the electrodes in a direction parallel to the electrode surface, in the plane of the diagram.   The mirrors are mounted so that radiation intensity and lasing action build up in a direction perpendicular to the plane of the figure.

The gas between the electrodes is initially preionized by placing a sharp pulse of high voltage on a row of spikes along the edge of the electrodes as shown in Figure 1.   High voltage on the spikes produces corona and ultra violet radiation at the tip of the spikes. This ultra violet radiation photoionizes the gas.

Immediately following this voltage pulse on the spikes, a voltage is applied to the planar electrodes.   Both voltage pulses are applied by transferring the voltage on a condenson bank by means of a spark gap.

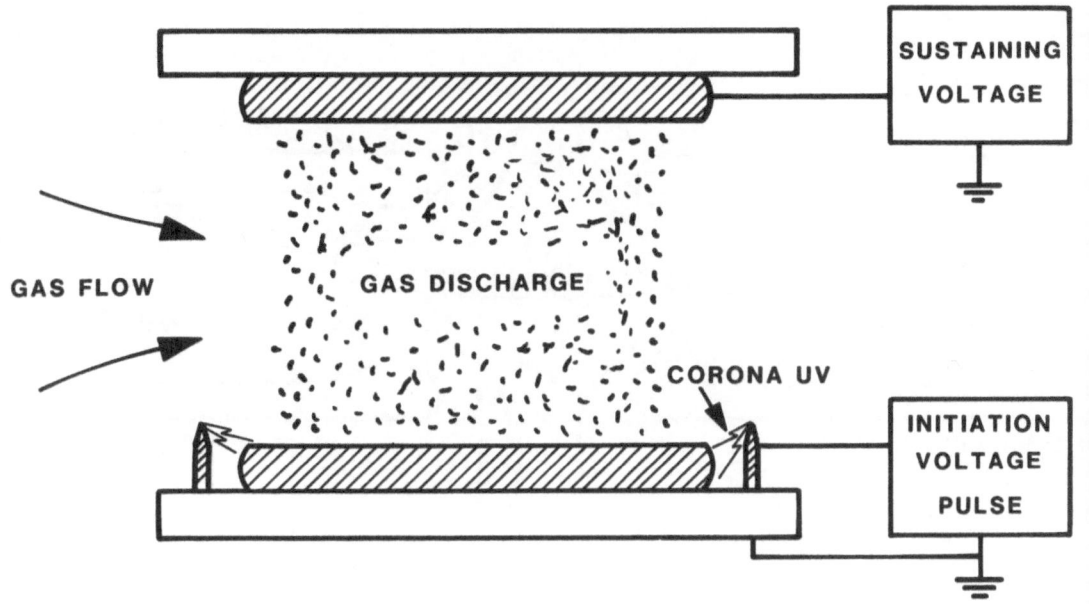

FIGURE 1 : Electrode configuration of a gas discharge laser.

The voltage across the planar electrodes, after an initial transient periods, settles down to a steady value.   This voltage is independent of current and is a characteristic of the laser gas mixture rather than the power supply.   Of course the voltage of the power supply must be sufficient to provide this voltage or the electrical discharge will not lase.

The problem to be solved is to determine, for a given gas mixture and electrode spacing, the magnitude of this steady voltage.

(2) Gas Blast Circuit Breakers -

To protect the generators of power stations, circuit breakers are required which can be the size of a large room.   The heart of the breaker is shown diagramatically in Figure 2. Two contacts, shown as cylindrical metal nozzles, are initially in contact, and then are separated to the position shown in Figure 2.   At the same time, as the contacts move apart, valves are opened to produce a high speed gas flow from a high pressure region outside of the nozzles, down the centre of the two nozzles.

On separation of the contacts an arc forms, initially, for example, at the two points A and B of Figure 2.   The gas flow, however, forces the arc to the centre of the nozzles, as shown in the figure.   The pressure driving the flow can be 20 atmospheres in the region outside of the nozzles and 1 atmosphere within the nozzles, so that the flow reaches supersonic velocities.   The currents that are required to be interrupted can be up to 50 kA. The transmission voltages are 330 kV for New South Wales, but are being upgraded to 500 kV.

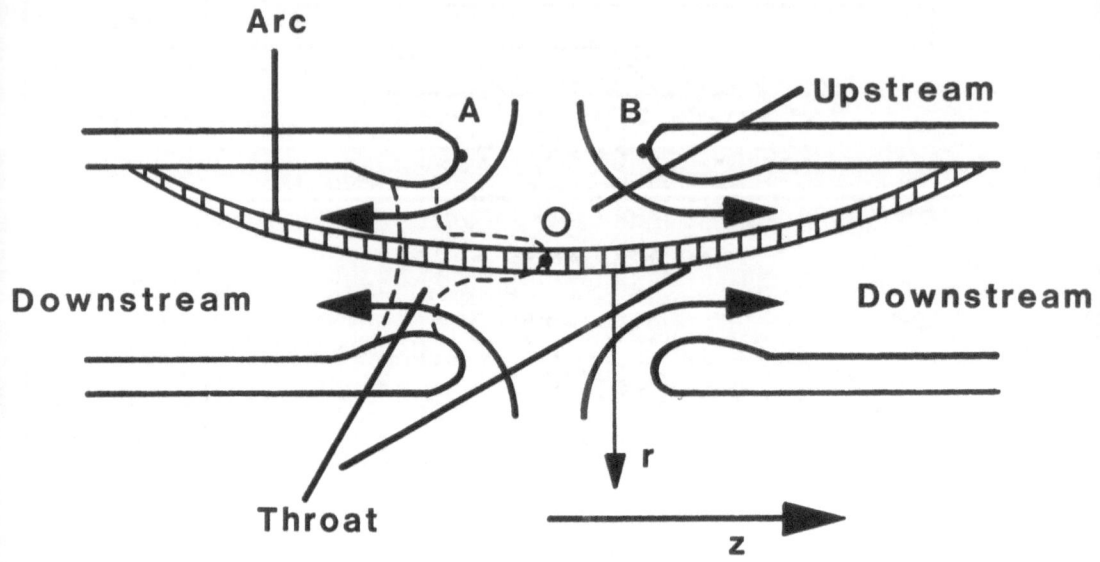

FIGURE 2 :   Simplified electrode configuration of a gas blast circuit ·breaker.

As the use of electric power increases the currents that are required to be interrupted become larger.   The larger the current, the larger is the diameter of the arc that is produced in the circuit breaker.   If the diameter of the arc is too large, it fills the entire nozzle. Then not only is there damage to the nozzle wall, but the circuit breaker is likely to fail to interrupt the required current, causing catastrophic damage to generating equipment.

The physical problem is to obtain an estimate of the arc diameter as a function of arc current and axial position in the nozzle.

3.   FORMULATION OF EQUATIONS

The most difficult task in the application of mathematics to industry is to decide which equations to solve.   Inevitably, approximations are necessary and a knowledge or insight of the physics of the dominating physical processes is essential.

(1) CO_2 lasers -

The gas laser, like the fluorescent lamp, is an example of a glow discharge.   In a glow discharge, it is known that the electrons have high energies, usually equivalent to a temperature of about 10 000 K, whereas the gas through which the electrons flow is close to room temperature.   The electrons, in the space of less than a millimetre, achieve an equilibrium energy distribution where the gain in energy from the applied electric field is balanced by energy losses due to collisions with the background gas.

For a fluorescent tube, which is at a pressure of about 0.01 atmosphere, the operating voltage is determined by the requirement that the loss of electrons to the wall by diffusion, must be equal to the rate of production of electrons by ionization. For CO_2 lasers however, a different process must be dominant because the pressures and dimensions are orders of magnitude larger so that diffusion cannot be the dominant electron loss process.

The solution to the problem depends on the physical insight that the electron loss process for CO_2 lasers is not diffusion, but electron attachment to CO_2 to form negative ions. Thus the process

$$e + CO_2 \rightarrow CO + O^-$$

occurs at low electron energies. At high electron energies, ionization is the dominant process. Thus the process is

$$e + CO_2 \rightarrow CO_2^+ + 2e$$

At different applied voltages; the electrons in the glow discharge have different average electron energies. The glow discharge operates at a voltage for which the average electron energy is such that the rate of electron loss by attachment is equal to the rate of electron production by ionization.

At any given electric field, the equilibrium distribution of electrons as a function of energy is determined by the Boltzmann transport equation. This equation is a type of energy balance equation relating energy gains from the electric field with energy losses due to elastic collisions as well as inelastic collisions with all the components of the gas mixture. The inelastic collisions are of many types - those contributing to rotational, vibrational and electronic excitation of the gas molecules, as well as ionization and attachment. The form of the Boltzmann equation that is solved has undergone development for a century (Huxley and Crompton [3]) and for a pure gas is

$$\frac{E^2}{3} \frac{d}{du}\left(\frac{u}{NQ_m} \frac{df}{du}\right) + \frac{2m}{M} \frac{d}{du}(u^2 NQ_m f) + \frac{2mkT}{Me} \frac{d}{du}\left(u^2 NQ_m \frac{df}{du}\right)$$

$$\tag{1}$$

$$+ \sum_j (u+u_j) f(u+u_j) NQ_j(u+u_j) - uf(u)N \sum_j Q_j(u) = 0$$

Here E is the electric field, N the gas number density, e and m are the electron charge and mass, respectively, M is the mass of a molecule of the gas, k is Boltzmann's constant, and u is the electron energy in volts. Thus, $u = mv^2/2e$ where v is the electron speed. The elastic cross section is Q_m, and Q_j is the cross section for the inelastic loss from the jth inelastic process.

The first term of Eq. (1) accounts for the gain in energy due to the electric field
E .   The second term, involving the factor 2m/M , accounts for elastic energy losses due
to collisions with the gas molecules.   The third term, involving the temperature T , enables
account to be taken of the gain in energy of low-energy electrons which collide with fast
moving thermal gas molecules.   The first term involving a summation over all values of j
accounts for electrons of energy $(u+u_j)$ losing energy u_j .   The second term in Q_j accounts
for electrons of energy u losing energy u_j .

To solve equation (1) for the distribution function f as a function of energy, the
cross sections Q_m and Q_j need to be known as a function of energy.   These cross-sections,
which for the 3 gases number 35 in all, are all available in the literature.   Incidentally
all were derived using experimental results obtained in Australia.

Once the distribution function has been evaluated for a given electric field, the
rate of ionization per electron or the rate of attachment, is determined by an integral of the
distribution function over the ionization or attachment cross section Q_j .   It is
conventional to express this rate of ionization or attachment by a coefficient α or a
defined by the relation

$$\frac{\alpha}{N} = \frac{8\pi e^2}{Wm^2} \sum_j \int_0^\infty Q_j f du \tag{2}$$

where the summation is made over the inelastic cross sections producing electrons for the
case of ionization, or over cross-sections for attaching processes for the case of attachment.
The electron drift velocity W is given by

$$W = \frac{E}{N} \frac{8\pi e^2}{3m^2} \int_0^\infty \frac{u}{Q_m} \frac{\partial f}{\partial u} du$$

(2) Circuit Breakers -

The physics of electric arcs is really a form of heat transfer theory.   Electrical
input to the arc plasma in the form of ohmic heating is balanced by heat losses due to
conduction, convection and radiation.   An energy balance equation expressing this relationship
defines arc temperature as a function of axial position and radius for an axisymmetric arc.
The problem is that this equation, while soluble, is too unwieldy for most practical purposes.

Complexity arises mainly from the radiation term (Lowke [5]).   The net emission or
absorption of radiation energy U per unit volume at any position is determined from the
divergence of the radiation flux $\vec{F}_R$ and thus $U = \nabla \cdot \vec{F}_R$.   The net radiation flux depends on
an integral of the radiation intensity I over all directions defined by angle Ω and unit
vector $\vec{n}$.   Thus $F_R = \int I \vec{n} d\Omega$.   Furthermore the radiation intensity in a given direction is
a sum over all wavelengths λ so that $I = \int I_\lambda d\lambda$ where I_λ is the radiation intensity per

unit wavelength at a particular wavelength λ. Finally I_λ, the radiation intensity in a particular direction and wavelength, depends on an integral along a line of sight, accounting not only for plasma emission along the line of sight, but also for absorption of radiation. Thus I_λ is given by the equation of radiative transfer familiar to astrophysicists, which is

$$\vec{n}.\nabla I_\lambda = \varepsilon_\lambda - I_\lambda K_\lambda \; ;$$

ε_λ is the radiation emission coefficient and K_λ is the absorption coefficient, which are known or at least can be estimated for a given plasma as a function of temperature, wavelength and pressure.

Many calculations have been performed for idealized situations solving these equations both exactly and using various approximations. Comparison of exact and approximate calculations has revealed an unexpected approximation which simplifies the calculations enormously. At the high currents that occur in circuit breakers, the central arc temperatures are 20 000 K or more. At these temperatures, most of the arc radiation is in the ultra violet portion of the spectrum. In Figure 3 are shown calculated radiation intensities for a typical high current arc, compared with black body radiation, indicated by broken curves. A significant fraction of this radiation has photons of sufficient energy to photoionize neutral gas atoms and molecules.

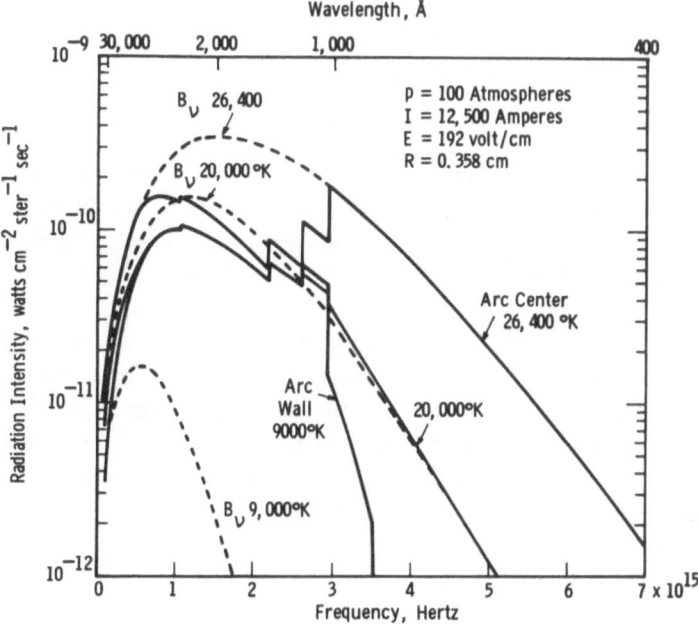

FIGURE 3 : Calculated radiation intensities for a typical arc compared with black-body radiation.

The basic approximation is to assume that all radiation is absorbed at the outer edge of the arc, helping to produce more arc plasma.   Thus the input electrical energy is assumed to be expended entirely in producing new arc plasma, which is swept beyond the downstream electrode in the imposed gas flow.   This energy balance is expressed by the relation

$$\sigma E^2 A = \partial(\rho h v_z A)/\partial z \; ; \tag{3}$$

σ is the electrical conductivity, E the electric field, A the arc area, ρ the plasma density, h the plasma enthalpy, v_z velocity of the plasma and z the axial distance from the stagnation point O midway between the electrodes of Figure 2.   The arc is assumed to be symmetrical about this point.

In equation (3) the first term gives the electrical input energy per unit length and $\rho h v_z A$ is the flux of the energy of the arc plasma at a particular axial position.   This flux is increased by the input electrical energy.

Unlike the more detailed calculations as a function of arc radius which preceded the formulation of the approximate model, the arc in equation (3) is represented by a simple channel of arc area A as a function only of the one dimensional variable z .   The material function σ , ρ and h need to be known at least approximately as a function of temperature.

A further equation linking E with arc current is Ohms law which is

$$I = \sigma E A \tag{4}$$

4.   SOLUTION OF EQUATIONS

1. Gas Lasers

Equation 3 is a second order differential equation with f as the only unknown as a function of energy u .   The equation has been solved by various methods but the most popular method of solution utilizes the particular form of the equation and has come to be known as the method of backward prolongation (Frost and Phelps [2]).

The equation is integrated with respect to u from o to u , utilizing the fact that the inelastic cross-sections are zero at zero energy.   Then a first order equation is obtained in df/du .   By setting f at an arbitrary value at a high energy such that values of f at higher energy are inconsequential, it is possible to evaluate f by extrapolating to lower energies, utilizing the numerical values of df/du .   The absolute value of f is obtained by normalization so that

$$\int 4\pi v^2 f \, dv = 1 \tag{5}$$

Typical numerical values of f(u) are given in Figure 4 for various value of the parameter E/N , the ratio of the electric field to the gas number density, for a particular ratio of CO_2 : N_2 : He of 1:2:3 , which is a common laser mixture (Lowke *et al* [4]).

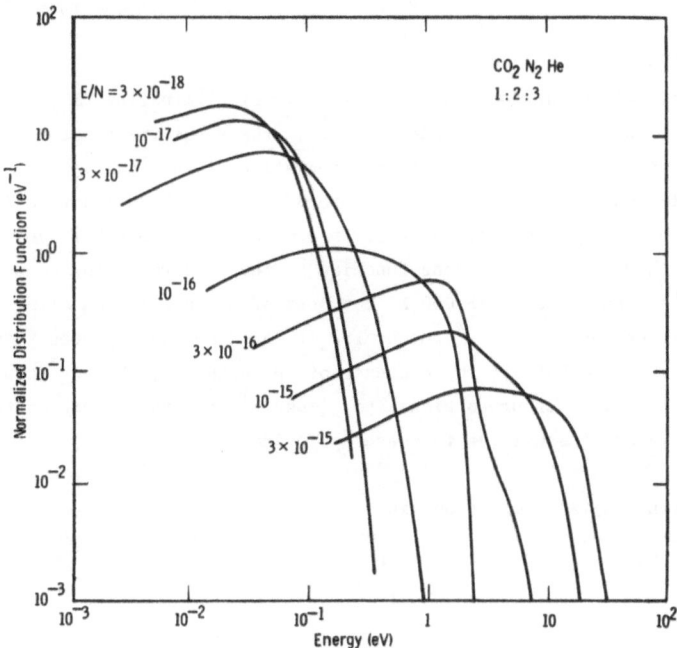

FIGURE 4 :   Example of calculated electron distribution function for CO_2 laser
mixture for various values of $E/N(V\ cm^2)$.

Knowing values of f , ionization and attachment coefficients α and a can then
be derived as shown in Figure 5.   At low electric fields, attachment dominates over ionization,
whereas at high electric fields, it is ionization which is dominant.   The gas laser operates
at a value of E/N for which ionization rates and attachment rates are equal, i.e. at the
value of E/N for which the curves for ionization and attachment intersect.

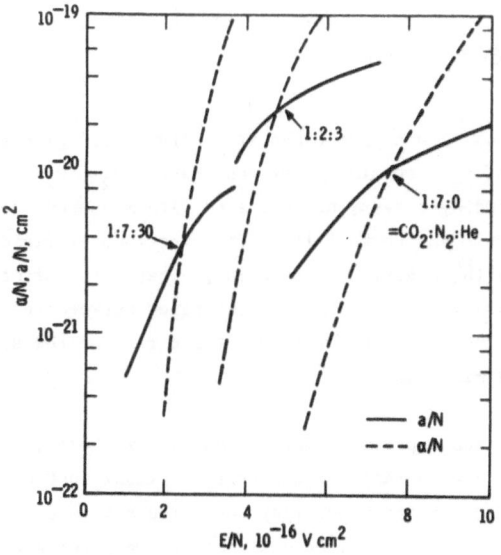

FIGURE 5 :   Calculated values of
ionization and attachment
coefficients for various CO_2
laser mixtures.

2. Circuit Breakers

Equations (3) and (4) can be developed analytically by eliminating E .   If it is
also assumed that $\rho h v_z$ and σ are independent of z , the resulting equation can be
integrated with respect to z .   The constant of integration is zero as $v_z = 0$ at the
stagnation point 0 of Figure 2, where we take z to be zero.   The product $\rho h/p$ is
approximately independent of z because, for a given pressure p , $\rho \propto 1/T$ and $h \propto T$
for a perfect gas.   Although σ is a strong function of temperature of low temperatures,
the temperatures will be greater than 15 000 K for most of the arc column along the nozzle,
so that σ is nearly independent of z .   Generally it would not be expected that the
increase of v_z with z would balance the decrease of p with z , but we actually obtain
a final expression for arc diameter proportional to $[\sigma\rho h v_z]^{\frac{1}{4}}$ and the effect of the $\frac{1}{4}$ power
is to make predictions rather insensitive to these approximations.

Thus from equations (3) and (4) we obtain

$$A = \left(\frac{2z}{\sigma\rho h v_z}\right)^{\frac{1}{2}} I \qquad\qquad (6)$$

Usually it is of particular interest to know the arc diameter at the nozzle throat, in which
case, for critical flow, we know that v_z is sonic velocity.   From values of the material
functions of air we set $\sigma = 100$ s/cm   and $\rho h/p = 8 Jm^{-3}Pa^{-1}$.   Then the arc diameter at
the nozzle throat is given by D in cm where

$$D = 0.5 \ I^{\frac{1}{2}} \ z^{\frac{1}{4}} \ p^{-\frac{1}{4}} \qquad\qquad (7)$$

where z is the distance from the upstream stagnation point in cm , p is the pressure in
bar at the throat of the nozzle and I is the current in kA .

5. INTERPRETATION

1. CO_2 Gas Lasers

The calculated values of E/N for which $\alpha = a$ are taken from Figure 5 to give the
operating electric field E for gas lasers of a given mixture of the gases CO_2 , N_2 and
He .   The calculated values of E , indicated by curves, are compared with measured values
indicated by points, as shown in Figure 6 (Denes and Lowke [1]).   The experimental fields are
obtained by dividing the operating voltage by the electrode separation, because the effect of
the electrodes on the voltage is known to be small.   The slight discrepancy between theory
and experiment is generally accounted for by including in the theory the effect of the small
amount of water vapour known to be in the gas mixtures.

There is good agreement between theory and experiment for various gas mixtures,
electrode separations and total gas pressures over a range of currents of almost four orders
of magnitude.   This agreement confirms the assumptions underlying the calculations and has
further consequences.   For optimum laser efficiency it may be desirable to vary the electric

field at which the laser operates.   From our understanding of the dominant physical processes, there is the facility to make such a change by adding a minority gas to vary α or a , and thus vary the value of E for which α = a .   For example one may add an easily ionizable impurity to increase α , or add a gas such as hydrogen which is known to detach O^- ions to attempt to reduce 'a' .

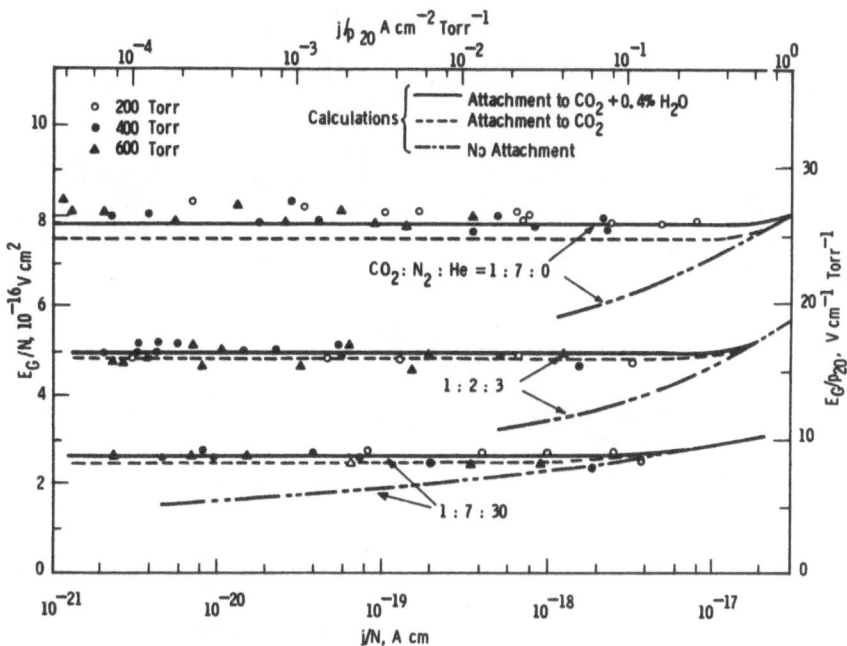

FIGURE 6 :   Comparison of theoretical predictions, given by curves, with experimental results, given by points, of operating electric fields in CO_2 lasers using various mixture rates.

2.   Circuit Breakers

In Figure 7 we show a comparison of theoretical values of the arc diameter at the nozzle throat as obtained from equation 7 with the experimental values, shown as points. Considering all the approximations that have been made, and the difficulties in making experimental measurements, the agreement is good.

Thus equation 6 and 7 enable circuit breaker engineers to make at least a first order estimate of arc diameters for various currents, so that ablation and damage of the nozzle of the circuit breaker can be avoided.

14

The utility of this simple physical picture of arc processes has been extended considerably further.   Equation 3 and 4 have been used to make predictions of arc temperatures, diameters and electric fields as a function of z for various currents and circuit breaker nozzles.   Agreement with detailed experimental results that have been taken for a 2000 A arc in nitrogen is very good.   These predictions using equations 3 and 4 avoid the many further assumptions required to develop equation 6 and 7.

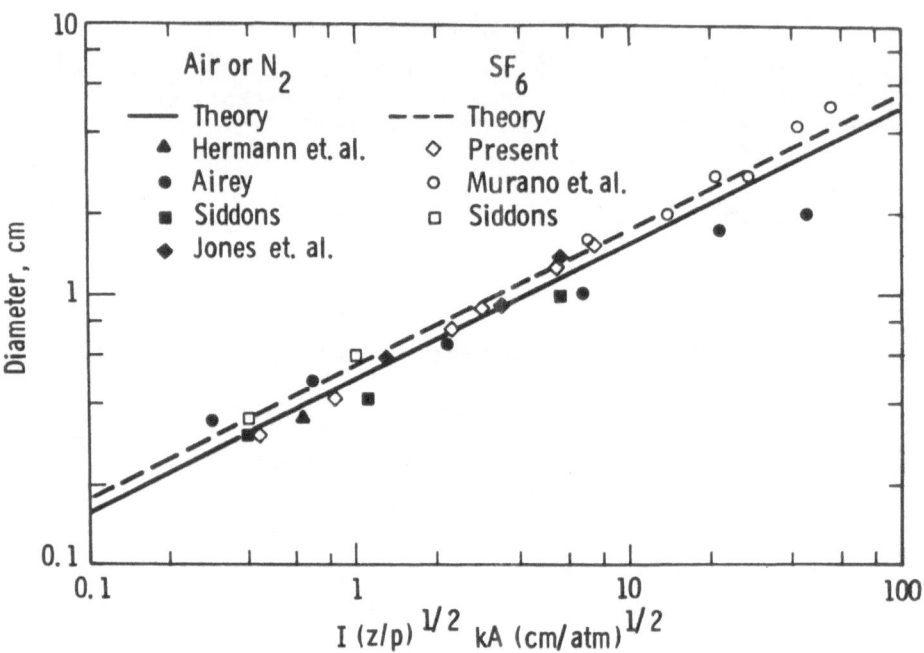

FIGURE 7 :   Comparison of theoretical predictions of arc diameter at the nozzle throat obtained from equations (6) and (7), with experimental results, shown as points.

6. REFERENCES

[1]   Denes, L.J. and Lowke, J.J. (1973)   V-I characteristics of pulsed CO_2 laser discharges, *Appl. Phys. Letters*, 23 (1973), 130.

[2]   Frost, L.S. and Phelps, A.V. (1962)   Rotational excitation and momentum transfer cross sections for electrons in H_2 and N_2 , *Phys. Rev.* 127 (1962), 1621-1633.

[3]   Huxley, L.G.H. and Crompton, R.W. (1974) *The diffusion and drift of electrons in gases*, Wiley, New York 1974.

[4]   Lowke, J.J. Phelps, A.V. and Irwin, B.W. (1973)   Predicted electron transport coefficients and operating characteristics of CO_2-N_2-He laser mixtures, *J. Appl. Phys.* 44 (1973), 4664.

[5]   Lowke, J.J. (1978)   Radiation transfer in circuit breaker arcs, in *Current interruption in high voltage networks*, edited by K. Ragaller, Plenum Publishing Co., New York, 1978, pp.299-327.

COMPOSITE REINFORCEMENT OF
CRACKED AIRCRAFT COMPONENTS

R. JONES AND A.A. BAKER,

Aeronautical Research Laboratories,
Department of Defence, Melbourne.

SUMMARY

This paper describes some work leading to the recent application of BFRP Crack-Patching to the field repair of fatigue-cracks in the aluminium alloy wing-skins of Mirage III aircraft.   Aspects covered include finite-element design procedures, and fatigue-crack propagation studies on patched panels, simulating the cracked and repaired area.   Repairs are currently being carried out by specially trained RAAF personnel during routine maintenance of the aircraft.

1.   INTRODUCTION

ARL (Australia) has pioneered the use of adhesively bonded boron-fibre reinforced plastic (BFRP) patches to repair cracks in aircraft components.   This procedure has been successfully used in several applications on RAAF aircraft, including the field repair of stress-corrosion cracks in the wings of Hercules aircraft and fatigue cracks in the landing wheels of Macchi aircraft [1,2].   Repairs were made by adhesively bonding the BFRP patch to the component with the fibres spanning the crack; the aim was to restrict opening of the crack under load, thereby reducing stress-intensity and thus preventing further crack propagation.   Adhesive bonding provides efficient load transfer into the patch from the cracked component and eliminates the need for the additional fastener holes (which introduce dangers such as fretting) associated with conventional mechanical repair procedures.   The advantages of using BFRP for the patch material include high directional stiffness (which enables reinforcement only in desired directions), good resistance to damage by cyclic loading and corrosion, and excellent formability.   Both BFRP and CFRP (carbon-fibre reinforced plastic) have the above advantages for use as patch materials, but BFRP was preferred for most practical applications because of its better combination of fatigue strength and stiffness and its higher expansion coefficient (which reduces the severity of residual stress on cooling following adhesive-curing at elevated temperature).   The low electrical conductivity of BFRP is also a very important advantage since conventional eddy-current procedures can then be used to detect and monitor cracks beneath the patch.

Recently, fatigue cracks were discovered in the lower wing skins, close to the main spar of some Australian Mirage aircraft, Figs. 1 and 2.   It was decided, in consultation with RAAF, that BFRP-patching would be an effective solution for this problem, because the repair would (i) cause no mechanical damage to the skin (i.e. no fastener holes), (ii) cause no strain elevation in the spar, since reinforcement need only occur across the crack, (iii) allow the use of conventional eddy-current procedures to check for crack growth, and (iv) allow implementation in the field during normal servicing, thereby minimising unavailability of the aircraft.   Because of the significance of the cracking and the long life desired of the repair an accurate mathematical model for adhesively bonded repair schemes was required.

We begin by discussing this model and then turn our attention to the design of the repair to the Mirage wing skin.   This is then followed by a brief description of the laboratory tests which were used to simulate the repaired structure.

At this point it should be mentioned that at the initial design stage it is much quicker and cheaper to design the repair using a digital computer than via a series of laboratory tests.   Indeed it is far easier to simulate the stress state in the structure numerically than experimentally.

2. BASIC THEORY

In developing fibre composite repair schemes (i.e. patches) for application to aircraft it is particularly important that a realistic model be used for the shear stress developed in the adhesive layer bonding the patch to the structure.   The model developed by the authors in [3,7] is given below for the specific case when the transverse shear moduli of the patch are equal.

As in most finite element methods we must first assume a functional form for the variation of the x,y displacements, which we will denote as u and v .   In this case we assume, in the case of a patch on one side of the structure only, that

$$u = u_o + \tau_{sx}(z(t_s+t_a+t_o-z/2) - (t_s+t_a+t_o/2)(t_s/2+t_a/2+3t_o/4))/G_o t_o \quad \text{in the patch}$$

$$= (u_o-3\tau_{sx}t_o/8G_o)(z-t_s)/t_a + (u_s-3\tau_{sx}t_s/8G_s)(t_s+t_a-z)/t_a \quad \text{in the adhesive}$$

$$= u_o + (z^2-t_o^2/4)\tau_{sx}/G_s t_s \quad \text{in the sheet} \tag{1}$$

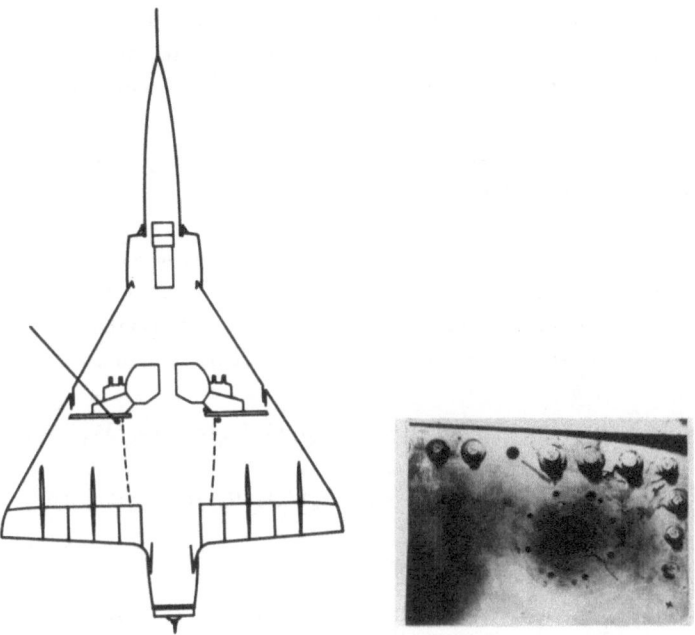

FIG. 1   Silhouette of Mirage III aircraft showing where the fatigue cracks developed in some
          aircraft.
          INSERT : Fuel decant hole region, showing the nature of the fatigue cracking.

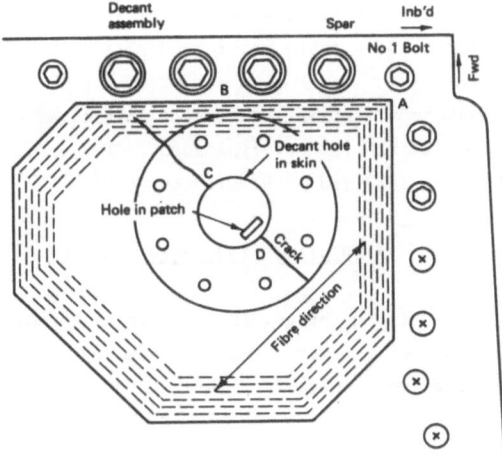

FIG. 2   Schematic diagram of fuel decant hole region with outline of BFRP patch superimposed;
          note the internal stepping of the patch.

and a similar expression for $v(x,y,z)$. Here G_o, G_s and G_a are the transverse shear moduli of the patch, sheet and adhesive respectively and t_o, t_s, t_a are their respective thicknesses. The x-y axis is taken in plane parallel to the mid-surface of the sheet and the z-axis is in the thickness direction with $z = 0$ being the equation of unpatched surface. The terms u_o, v_o and u_s, v_s refer to the x and y movements at the mid-surface of the patch and sheet respectively while τ_{sx} and τ_{sy} are the adhesive shear stresses in the x, and y directions.

This form for $u(x,y,z)$ and $v(x,y,z)$ gives a continuous displacement field through the thickness of the sheet, adhesive and patch, and gives zero shear stresses on the free surfaces of the patch and the sheet. With this formulation the relationships between displacements and shear stresses in the adhesive layer reduce to :

$$\tau_{sx} = (u_o - u_s)/(t_a/G_a + 3t_s/8G_s + 3t_o/8G_{13}) \tag{2}$$

$$\tau_{sy} = (v_o - v_s)/(t_a/G_a + 3t_s/8G_s + 3t_o/8G_{23}) \tag{3}$$

which is similar to the relationship developed in [4].

The terms $\left[\dfrac{t_a}{G_a} + \dfrac{3t_s}{8G_s} + \dfrac{3t_o}{8G_o}\right]$ and $\left[\dfrac{t_a}{G_a} + \dfrac{3t_s}{8G_s} + \dfrac{3t_o}{8G_o}\right]$ may be regarded as spring

constants and for typical repairs differ by 20-50% from the simplistic approximation of G_a/t_a used by other investigators. (See [3] for further details). The effect of this error on the stiffness matrix for the structure is to overestimate the stiffness of the adhesive layer. However, as shown in [3,7], this has little effect on the efficiency of the patch, (e.g. on the reduction in the stress intensity factor). But the simplistic model drastically over-estimates the shear stress in the adhesive. This is not important in lightly loaded areas, but for repairs to areas which carry a substantial amount of load it would often mean the unnecessary rejection of the concept of a fibre reinforced plastic patch.

This point is particularly important in the present investigation since the wing skin in the vicinity of the spar and root rib is heavily loaded. For experimental work of direct relevance to the above, readers are referred to References [4,9].

3. ADHESIVE STIFFNESS MATRIX

In the previous section we derived the governing expressions for the transverse shear stresses τ_{xz} and τ_{yz} in the adhesive layer. We will now concentrate on developing the stiffness matrix for an adhesive element where, for the sake of simplicity, we will assume that bending does not occur.

The strain energy V_a of an element of the adhesive is given by :

$$V_a = \frac{1}{2G_a} \int_{t_s}^{t_s+t_a} \iint_e (\tau_{zx}^2 + \tau_{zy}^2) \, dxdydz = \frac{t_a}{2G_a} \iint (\tau_{zx}^2 + \tau_{zy}^2) \, dxdy \tag{4}$$

where the double integration is over the area of the element. The element stiffness matrix, K_a^e for the adhesive is now obtained by first prescribing the form of the shear stress variation within the element and then by differentiating the strain energy with respect to each of the elemental degrees of freedom. However before prescribing the nature of the shear stresses we must first stipulate the geometry of the element to be considered. Here we will confine our attentions to triangular elements which are the basic elements from which all other elements may be assembled. Let us assume that, within the element, the shear stresses vary linearly with x and y , this is the same as assuming a linear displacement field, i.e.

$$\tau_{zx} = [(a_i+b_ix+c_iy)\tau_{zxi} + (a_j+b_jx+c_jy)\tau_{sxj} + (a_m+b_mx+c_my)\tau_{zxm}]/2\Delta \tag{5}$$

where

$$a_i = x_jy_m - x_my_j; \quad b_i = x_m - x_j; \quad c_i = x_m - x_j \tag{6}$$

and a_j, a_m, b_j, b_m and c_j, c_m are obtained by a cyclic permutation of the indices i,j,m . Here (x_i,y_i) , (x_j,y_j) and (x_m,y_m) are the coordinates of the corners of the element, Δ is the area of the element and $\tau_{zxi}, \tau_{zxj}, \tau_{zxm}$ are the values of the shear stresses at the nodes i,j and m respectively. Hence if we now substitute for τ_{zxi}, τ_{zxj}, and τ_{zxm} as given by equations (2) and (3) we obtain :

$$\tau_{zx} = \sum_{p=i,j,m} \phi_p(x,y) F(u_o-u_s)p \tag{7}$$

and

$$\tau_{yz} = \sum_{p=i,j,m} \phi_p(x,y) F(v_o-v_s)p \tag{8}$$

where $\phi_i(x,y) = (a_i+b_ix+c_iy)/2\Delta$ and where p takes on the values, i,j , or m . Substituting for τ_{xz} , and τ_{yz} into equation (4) we obtain the following expression for the strain energy :

$$V_a = \frac{t_a}{2G_a} \iint [F\{\Sigma_p\phi_p(x,y)(u_o-u_s)_p\}^2 + F\{\Sigma_p\phi_p(x,y)(v_o-v_s)_p\}^2] dxdy \tag{9}$$

Here p and q are dummy indices which take on the node values of either i,j or m and where, for the sake of convenience, we have written :

$$F = 1/(t_a/G_a + 3t_s/8G_s + 3/8G_o) \tag{10}$$

Equation (9) directly relates the strain energy of the adhesive to the displacements in the sheet and the patch and clearly illustrates the importance of the previous section where the exact nature of the shear stress in the adhesive was determined.

The element stiffness matrix is now obtained by differentiating V_a with respect to each of the elemental degrees of freedom, i.e.

$$K_{a\sim}^e \lambda = \frac{\partial V_a}{\partial \lambda_{\sim}} \tag{11}$$

where $\lambda_{\sim}$ is the vector

$$\lambda_{\sim}^T = [u_{si}, v_{si}, u_{oi}, v_{oi}, u_{sj}, v_{sj}, u_{oj}, v_{oj}, u_{sm}, v_{sm}, u_{om}, v_{om}] \tag{12}$$

This technique is explained in detail in reference [3], where the full form of the 12×12 stiffness matrix is given. The stiffness matrix for the adhesive element is now used in conjunction with standard finite element routines for the sheet and the patch except at a crack tip where it is coupled to a special crack tip element. Let us now illustrate the use of this approach by considering the specific repair to the lower wing skin of Mirage III aircraft mentioned previously.

4. PATCH DESIGN AND ANALYSIS

The fatigue cracks initiated near the fuel decant hole in the lower wing skin, close to the intersection of the main spar and root rib (Fig.1). The skin consists of aluminium alloy AU4G, similar to 2014T6, about 3.5 mm thick and is covered by a fairing. The wing skin Panel is in a state of shear which results in the fatigue cracks propagating at 45° to the spar (Fig.2). For design purposes the maximum size of the crack (including the fuel decant hole) was taken as 111 mm. Flight loads were estimated from strain-gauge data from a wing fatigue test and from the manufacturer's stress analysis.

The BFRP patch (Fig.2) is a unidirectional laminate with fibres running perpendicular to the direction of crack propagation. The patch contains seven layers of BFRP and is internally stepped, i.e. the largest layer is on the outside to reduce inter-laminar shear stresses. The proximity of the spar bolts and the root rib bolts to the decant hole necessitated closer spacing of the layer steps in this region. The attachment holes for the decant hole cover were preformed in the patch; the decant hole opening was much reduced in size by the patch, so as to obtain maximum reinforcement efficiency.

The main objectives of the numerical procedure were to assess the reduction in stress-intensity in the skin, reveal any undue strain elevation in the spar and to estimate the maximum levels of stress in the BFRP patch and adhesive layer. Ideally, a sufficient reduction in stress-intensity should be obtained without exceeding the materials allowable in the patch system, the most critical of which is the adhesive shear strength under fatigue loading in the operating environment.

A detailed finite-element stress analysis of the cracked region was undertaken using standard procedures. The resulting stress-intensity factors, assuming loads typical of a 6g flight manoeuvre were $K_1 = 57$ MPa $\sqrt{m}$, $K_2 = 3$ MPa $\sqrt{m}$ at the tip nearest to the spar, and $K_1 = 54$ MPa $\sqrt{m}$, $K_2 = 1$ MPa $\sqrt{m}$ at the tip nearest to the root rib.

The effect of the patch was estimated by adding a finite-element representation of the BFRP patch to the crack model (Fig.3).   This representation consisted of 380 of the bonded elements previously described.

Several patch adhesive thickness combinations were considered each with the same plan form shown in Figs.2 and 3.   Finally, from the results of the analysis a seven-layer patch was chosen, maximum thickness about 0.89 mm, with an adhesive layer 0.2 mm thick.   For this patch, the stress-intensity factor (K_1) was reduced by at least 90% and the fibre strains were well below the maximum working level of 0.005 (equivalent to a stress of 1000 MPa, assuming a composite modulus of 200 GPa).   The shear stresses in the adhesive were (Fig.2) :

Point A	14 MPa	at edge of patch
Point B	24 MPa	at edge of patch
Point C	51 MPa	at edge of decant hole
Point D	41 MPa	at edge of decant hole

At locations C and D, the shear stress in the adhesive could cause localised fatigue damage [5] but more detailed analysis showed that this damage was only very localised and that small areas of debond in this region should not adversely affect the patch efficiency.

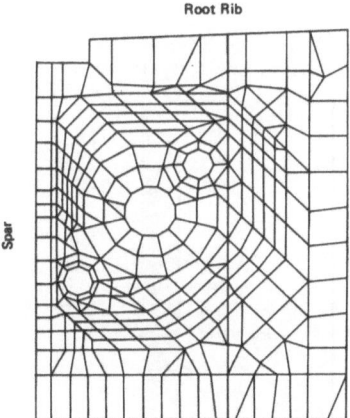

FIG. 3 :   Finite-element mesh for the fuel decant hole region.

The effect of the patch on the maximum principal stress in the skin along the line of attachment to the spar (Fig.4) showed that the patch effectively restores the stress distribution along the spar to that in a wing with an uncracked panel.   Consideration was also given to thermal and residual stress effects in the aluminium skin and to thermal-fatigue

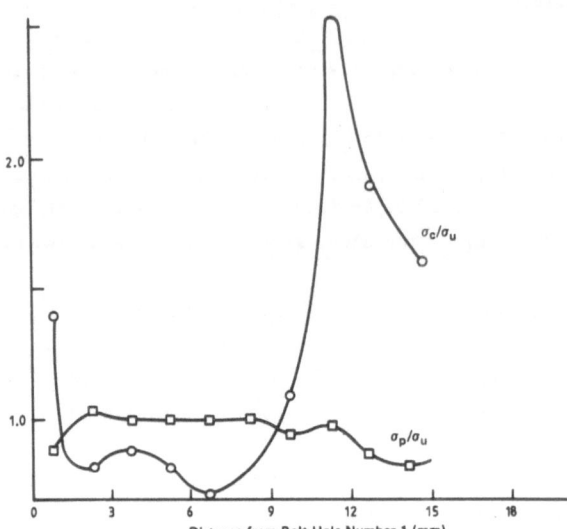

-FIG. 4 : Stress distribution in the aluminium skin.   The curves are for the stress,
         σ , for ratios between three cases denoted by subscripts; p = patched and
         cracked,   u = unpatched and uncracked, and c = cracked and unpatched.

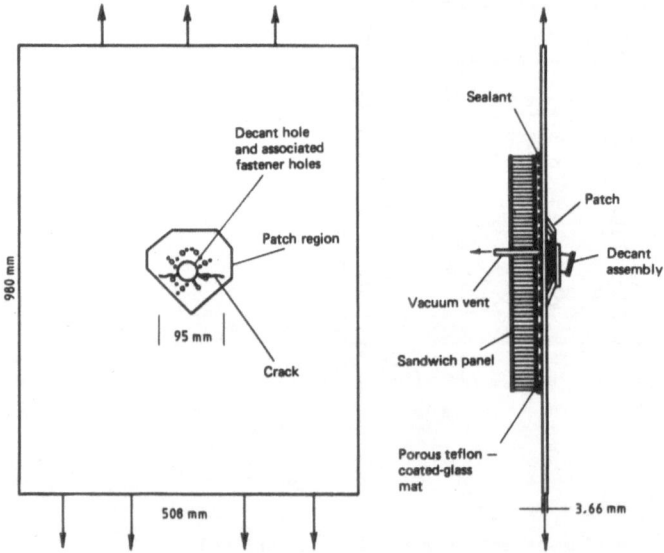

FIG. 5 : Details of 2024T3 aluminium alloy wing-skin simulation panels used for
         crack-propagation studies, showing position of patch and vacuum box restraint.

effects in the adhesive due to the use of an adhesive curing at elevated temperature (-120°C) and the mismatch in thermal expansion between the aluminium alloy and BFRP. It was concluded [6,8] that these effects posed no serious problems in this particular repair.

The BFRP repair was qualified mainly by two series of tests (a) a strain survey on a Mirage with a patch installed to show that the patch did not significantly elevate the local strain in the spar, and (b) fatigue tests on aluminium alloy panels configured to simulate the cracked area in the wing (Fig.5). The aims of the fatigue test were to check that (i) the predicted reduction in stress-intensity could be achieved, and (ii) the patch/adhesive system could endure the fatigue loading. The strain survey confirmed that no significant strain elevation occurs in the spar after patching. The fatigue results are given below.

5. FATIGUE CRACK PROPAGATION STUDIES

Tension-tension fatigue tests were undertaken on the aluminium alloy panels using either constant load cycling or block load cycling. Only the principal tensile stress in the wing could be simulated. The tests were carried out in a laboratory environment at a temperature of 23°C. Tests under more realistic environmental conditions are planned. Initially, due to unavailability of 2014T6 material, tests were carried out on aluminium alloy 2024T3 panels. Although this alloy has significantly greater resistance to crack propagation than the wing skin alloy AU4G, it was considered that these tests would show whether or not the stress-intensity was reduced by patching and if the fatigue characteristics of the adhesive layer were satisfactory. Cracks were initiated from saw-cuts and propagated to about 100 mm total length prior to BFRP patching, using the procedures described below. During application of the patch, the panels were constrained at their ends to simulate the constraint to thermal expansion that would be experienced in the wing-skin under practical repair conditions. This rate of crack growth was monitored during the initial test by direct observation with a travelling vernier microscope. After patching, eddy-current procedures were used to monitor the crack under the patch.

In the early tests local bending of patched panels occurred under load due to local displacement of the neutral axis of the panel by the patch. This effect negated the reinforcing effect of the patch to a large extent. Strain-gauge tests confirmed the presence of substantial strains in the panel on the opposite side to the patch. In the actual wing-skin, local stiffening by the surrounding structure, particularly the spar, root rib and internal stiffeners, would largely prevent bending and a method was therefore devised to minimize its occurrence in the test. This procedure used atmospheric pressure to hold the panel against a relatively stiff, flat honeycomb sandwich panel (by evacuating the region between the back of the test panel and the sandwich panel). A thin layer of porous, fluorocarbon-coated fibre-glass between the test panel and the sandwich panel acts to allow air removal in the gap between the test panel and sandwich panel and minimised the frictional load. Strain-gauge studies showed that the bending effects were then reduced by about 50%.

Using the vacuum restraint, substantial increases in life prior to crack growth were achieved; Fig.6 plots crack growth curves for three 2024T3 specimens subjected to constant load

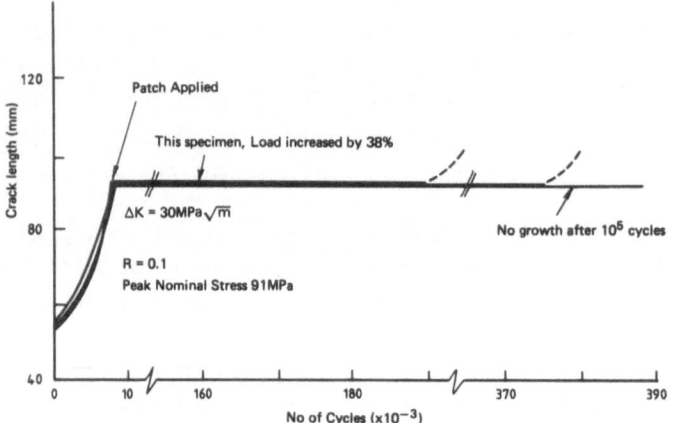

FIG. 6 : Crack length versus cycles for 2024T3 wing-skin simulation panels
         subjected to constant load cycling, stress intensity range prior to
         patching is indicated.

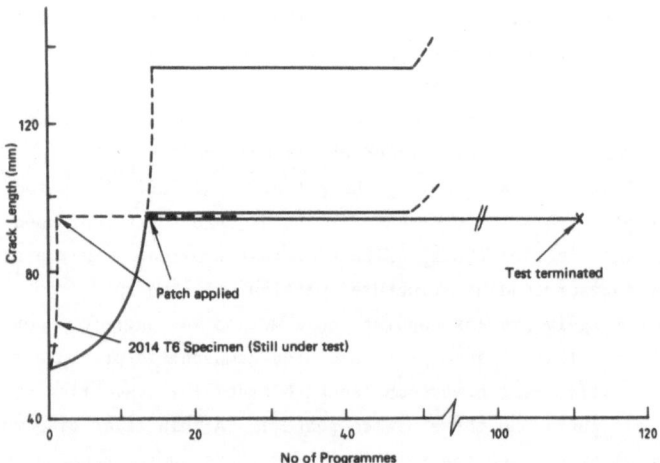

FIG. 7 : Crack length versus number of programs of block loading for 2024T3 and
         2014T6 wing-skin simulation panels.

cycling at an applied stress of 91 MPa, (equivalent to a nominal 4g manoeuvre load). In all cases initial crack growth rates indicated that failure of the panels (without the repair) would have occurred within a further 10^4 cycles.

Panels were tested under more realistic loading conditions using a block-loading sequence adapted from the Mirage load spectrum previously developed at ARL. Each program sequence is approximately equivalent to one year of aircraft usage. Substantial life extensions for the cracked 2024T3 alloy panels, (Fig.7) indicate that, provided local bending is minimised, the BFRP patch system provides a significant stress-intensity reduction. The actual life extension probably depends largely on the degree to which local bending is reduced in practice.

6. CONCLUSION

We have seen how simple mathematical concepts can be used to solve complex engineering problem. Indeed the field of "crack-patching" is one in which the inputs from the disciplines of mathematics, engineering and materials science have accelerated the development of a very cost effective technology.

The advantages of crack-patching over conventional repair procedures are :

a) no mechanical damage to surrounding structure, no fastener holes,

b) crack can be inspected through the patch by conventional eddy-current NDI,

c) the cracked area is protected from further external corrosion, and internal fuel leakage, by the patch system,

d) reinforcement is only in the direction required, no undesirable stiffening in other directions, and

e) patches can be removed and replaced with no damage to the surrounding structure.

7. REFERENCES

[1]   BAKER, A.A.,   Composites, Vol. 9, 1978, pp.11-16.

[2]   BAKER, A.A. and HUTCHINSON, M.M.,   "Fibre Composite Reinforcement of Cracked Aircraft Structures" Proceedings of 33rd Annual Technical Conference of the Plastics Industry, 1978, Section 17-E.

[3]   JONES, R. and CALLINAN, R.J.,   J. Struct. Mech. Vol. 7, 1979, pp.107-130.

[4]   MITCHELL, R.A., WOOLEY, R.M. and CHIVIRUT, D.J.,   A.I.A.A., Vol. 13, 1975, pp.744-749.

[5]   BAKER, A.A.,   S.A.M.P.E. Journal, Vol. 15, 1979, pp.10-17.

[6]   BAKER, A.A., DAVIS, M.J. and HAWKES, G.A.,   "Repair of Fatigue Cracked Aircraft Structures with Advanced Fibre Composites :   Residual Stress and Thermal Fatigue Studies", to be published in Proceedings of the 10th International Committee on Aeronautical Fatigue Symposium, May 1979.

[7]   JONES, R. and CALLINAN, R.J.,   J. Fibre Science and Technology, 1980, 14, 2, 99-111, 19

[8]   JONES, R. and CALLINAN, R.J.,   J. Struct. Mech., Vol. 8, 1980, pp.144-149.

[9]   BAKER, A.A., CALLINAN, R.J., DAVIS, M.J., JONES, R. and WILLIAMS, J.G.,   "Application of BFRP crack patching to Mirage III aircraft", Proceedings 3rd Int. Conf. Comp. Materials, August 1980, Paris, France (in press).

MATHEMATICAL MODELS FOR GAS DISTRIBUTION
IN THE IRONMAKING BLAST FURNACE

J.M. BURGESS, D.R. JENKINS,

Central Research Laboratories,
Broken Hill Proprietary Limited, Newcastle,

AND

F.R. DE HOOG,

Division of Mathematics and Statistics,
CSIRO, Canberra.

ABSTRACT

The distribution of gas flow through the ironmaking blast furnace is of fundamental importance in the control of furnace productivity, fuel rate and campaign life. Mathematical models which describe the flow distribution of gases have been developed for use in both fundamental and applied analysis of the blast furnace gas flow problem.

The gas flow distribution through the furnace softening-melting zone, which consists of a series of interconnected packed beds of coke particles separated by impervious ferrous layers, has been predicted by solution of a series of non-linear compressible flow pressure loss equations based on an empirical relationship by Ergun. A further model, which employs the vectorial form of the Ergun equation, has been developed for prediction of the two-dimensional flow of gas exiting from the softening-melting zone and its subsequent distribution through the stack of the furnace.

This paper describes the solution procedures employed in the above models and their application to actual blast furnace conditions. As well as providing fundamental understanding of blast furnace operations, the practical applications of these models include prediction of the softening-melting zone shape using blast furnace shell pressure tapping information; interpretations of furnace gas temperature and composition measurement probes located in the shaft; and determination of optimum conditions for improved gas distribution within the furnace.

INTRODUCTION

The Blast Furnace Process

The iron blast furnace is a packed bed reactor into which alternate layers of coke and
ferrous burden materials are charged at the top and a hot air blast is blown through ports at
the bottom, called tuyeres.   A schematic diagram of the zones formed within the blast furnace
is shown in Figure 1.   The hot air blast entering through the tuyeres reacts with lumps of
coke within the combustion zone, or raceway, to form reducing gases (carbon monoxide and
hydrogen).   This gas then flows upwards through the furnace to react with the iron oxide in
the ferrous layers, ultimately reducing the iron ores to molten iron.   The charged layers of
burden materials slowly descend into the furnace and, as they descend, the gas-solid contact
causes an increase in the temperature of the material due principally to convective heat
transfer.   The packed bed of ferrous materials retains its particulate or "lumpy" nature
until the temperature of the ferrous material reaches its softening temperature.   This region
of the blast furnace is consequently known as the lumpy zone.

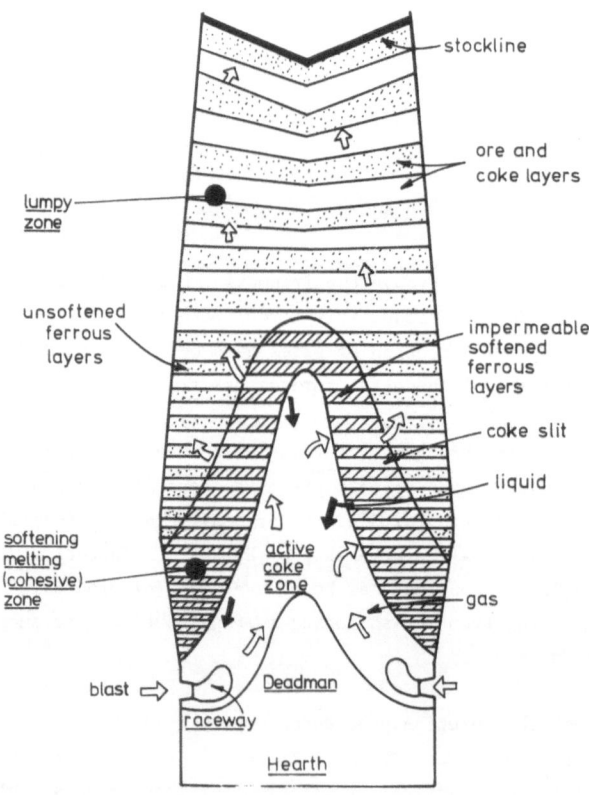

FIGURE 1 : Schematic of zones in the blast furnace.

Once the ferrous material softening temperature is reached, the ferrous material becomes a cohesive mass which has low permeability to gas flow. Because the burden materials are not charged in a uniform manner across the radius of the furnace, and also due to the preferential flow of solids towards the periphery of the furnace at the raceway zone, the depth in the furnace at which the ferrous materials soften is a function of the radius in the furnace. As the temperature increases further, the ferrous material melts and flows downwards through the furnace to the hearth. The softening and melting mechanism then results in a series of annular rings of softened material, separated by layers of coke. This region of the furnace is called the softening-melting zone of cohesive zone.

Due to the relatively impermeable nature of the softened layers, gas flowing upwards from the raceway must pass predominantly through the permeable coke layers, called "coke slits", between the softened materials, in order to reach the lumpy zone above. Consequently, the geometrical shape of the cohesive zone, along with its position in the furnace and the dimensions of the coke slits, is one of the major factors in determining the distribution of the flow of gas through the upper zone of the furnace.

The molten material flowing from the melting part of the cohesive zone passes through a packed bed of coke to the bottom of the furnace. This zone is called the hearth, and the liquids collect in this zone and are tapped periodically from the furnace. The liquid products consist of two components - the molten iron, or hot metal, and the slag. The slag is produced from the melting of the non-ferrous or gangue components of the burden material, and from the ash component of the coke which is released during combustion. The slag has a lower density and higher viscosity than the molten iron, although both liquids are tapped from the furnace simultaneously.

The BHP Co. Ltd. has 12 blast furnaces, 11 of which are currently in operation, at its four steelworks located in Port Kembla (5 blast furnaces), Newcastle (4), Whyalla (2) and Kwinana (1). The annual iron production from these furnaces is approximately 7.3 million tonnes, and the annual coke consumption is approximately 5.1 million tonnes. The largest of these blast furnaces is No.5 Port Kembla, which has an inner volume of 2993 m^3, 28 tuyeres and a hearth diameter of 12.15 m . The annual capacity of this furnace is approximately 2.1 million tonnes. In contrast, the smallest of the Company's blast furnaces is No.1 furnace at Newcastle Steelworks which has an inner volume of 636 m^3, 10 tuyeres and a hearth diameter of 5.49 m . The annual capacity of this furnace is approximately 340,000 tonnes.

In order to optimise blast furnace profitability, the operating conditions need to be such that the rate of coke consumption per tonne of iron is minimised, the life of the refractory lining of the furnace walls is maximised, and the production rate is optimised according to market conditions. Therefore, research into blast furnace operation must be directed towards these objectives. It is for these reasons that a programme of investigation into the fundamental aspects of the flow of gas within the blast furnace has been undertaken. This paper describes some of the mathematical models developed as part of this programme.

Analysis of Blast Furnace Operation

Much information relating to the internal state of blast furnaces has been obtained by Japanese workers through dissection of a number of quenched blast furnaces (Kanbara et al. [3]) Figure 2 illustrates the shape of cohesive zones found in three of these quenched furnaces.

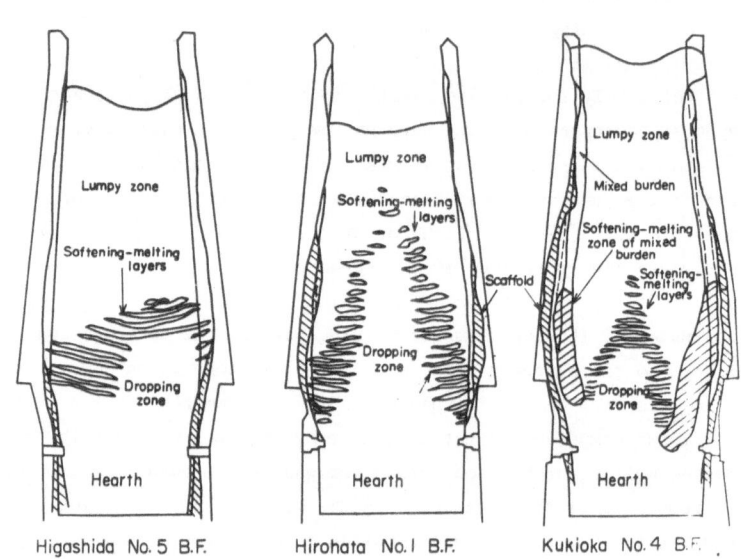

Higashida No. 5 B.F.   Hirohata No. 1 B.F.   Kukioka No. 4 B.F.

FIGURE 2 : Softening-melting zones found in quenched blast furnace (Kanbara et al. [3]).

Higashida No.5 blast furnace showed an extremely irregularly shaped cohesive zone, which was associated with irregular furnace operation.   The cohesive zone shape shown for Hirohata No.1 blast furnace, on the other hand, was a high, inverted V-shaped zone, and this furnace experienced low resistance to gas flow and high productivity during its operation.   Finally, Kukioka No.4 blast furnace exhibited a "W-shaped" cohesive zone with much thinner layers within the zone, leading to higher resistance to gas flow.   These three different samples of cohesive zones illustrate the variety of shapes which may occur in the blast furnace and give some insight into the relationships between cohesive zone shape and furnace operation.   It would therefore be desirable to be able to predict the shape and position of the cohesive zone within the furnace based on information obtained from furnace instrumentation.   Unfortunately the amount of useful information which can be obtained is limited by the harsh environmental conditions within the blast furnace and inaccessibility of suitable measurement points.   On a number of BHP blast furnaces, temperature probes are located above the surface of the burden to measure the radial distribution of temperature, and thermocouples are located in the cooling water system for the furnace walls.   Also, on three of the Company furnaces, a small number of pressure tapping points are located along a vertical profile of the furnace shell.

Finally, on No.5 blast furnace Port Kembla, a probe located approximately 4 metres below the stockline of the furnace may be pushed into the burden material to measure the radial variations in gas temperature and gas composition at this level. The relative locations of these measurement devices are illustrated in Figure 3.

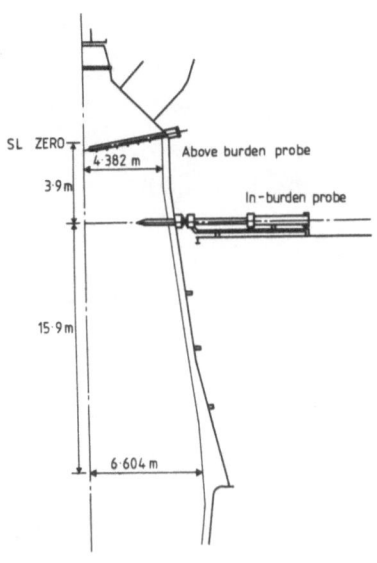

FIGURE 3 : Location of measurement devices on Port Kembla No.5 blast furnace.

In order to use these measurement devices for prediction of furnace condition, and specifically for prediction of cohesive zone geometry, it is necessary to develop mathematical models to describe processes occurring within the furnace which use measured furnace data as input and provide a suitable output.

GAS FLOW IN PACKED BEDS

The flow of gas within the iron blast furnace is an example of a countercurrent gas-solid reactor. The flow velocity of the solid material is much smaller than that of the gas, and for most considerations may be neglected. Thus, analysis of the gas flow within the blast furnace is obtained through use of the theory of the flow of gas in packed beds. The most commonly used equation for the pressure drop in such systems is the Ergun equation :

$$- \frac{\Delta P}{L} = \frac{150\mu v}{d_p^2} \frac{(1-\varepsilon)^2}{\varepsilon^3} + \frac{1.75\rho v^2}{d_p} \frac{(1-\varepsilon)}{\varepsilon^3} \tag{1}$$

where μ is the gas viscosity, ρ is the gas density, v is the superficial velocity of the gas, ε is the voidage of the packed bed and d_p is the mean particle diameter and the

constants 150 and 1.75 have been empirically determined.   A concise derivation of the Ergun
equation is provided by Bird *et al.* [1].   The Ergun equation has been generalised by a number
of workers (e.g. Stanek and Szekely [5]) to beds which are not uniformly packed due to
variations in particle size and voidage, which is the situation existing within a blast furnace
For such applications the Ergun equation may be written in the vector form

$$- \nabla P = (c_1 \mu + c_2 \rho |\underset{\sim}{v}|) \underset{\sim}{v} \tag{2}$$

where

$$c_1 = \frac{150}{(\phi_s d_p)^2} \frac{(1-\varepsilon)^2}{\varepsilon^3}$$

$$c_2 = \frac{1.75}{\phi_s d_p} \frac{(1-\varepsilon)}{\varepsilon^3}$$

and $|\underset{\sim}{v}|$ is the magnitude of the superficial velocity vector.   If the ideal gas law

$$\frac{P}{\rho T} = \frac{R}{M}$$

is assumed to hold, then (2) can be rewritten as

$$- \nabla P^2 = \frac{2TR}{M} (c_1 \mu + c_2 |\rho \underset{\sim}{v}|)(\rho v) \tag{3}$$

Finally, if no gas is generated in the bed and the flow is steady state, the equation of
continuity is

$$\nabla \cdot (\rho \underset{\sim}{v}) = 0 \tag{4}$$

Equations (3) and (4) can be simplified considerably when the flow is axially symmetric.   If
we take v_r and v_z to denote the radial and vertical components of the velocity $\underset{\sim}{v}$, and
define the Stokes stream function ψ by

$$\rho v_r = - \frac{1}{r} \frac{\partial \psi}{\partial z} , \qquad \rho v_z = \frac{1}{r} \frac{\partial \psi}{\partial r} ,$$

the continuity equation (4) is satisfied automatically.   Then, on taking the curl of equation
(3) and noting that $\nabla \times \nabla P^2 = 0$, we obtain

$$\frac{\partial}{\partial r}\left[\left(C\mu + \frac{1}{r} \sqrt{\left(\frac{\partial \psi}{\partial r}\right)^2 + \left(\frac{\partial \psi}{\partial z}\right)^2}\right) \frac{T}{r} \frac{\partial \psi}{\partial r}\right]$$

$$+ \frac{\partial}{\partial z}\left[\left(C\mu + \frac{1}{r} \sqrt{\left(\frac{\partial \psi}{\partial r}\right)^2 + \left(\frac{\partial \psi}{\partial z}\right)^2}\right) \frac{T}{r} \frac{\partial \psi}{\partial z}\right] = 0 \tag{5}$$

where

$$C = \frac{c_1}{c_2}$$

Equation (5) is an equation in the variables T, μ and ψ. To a good approximation, the viscosity μ is a function of the temperature T and the gas under consideration. Thus, if T is known as a function of r and z, (5) is a nonlinear elliptic partial differential equation of second order for the stream function ψ.

As is clear from Figure 1, the packed bed under consideration consists of layers of coke and ferrous materials and in each of these layers, the voidage ε, the shape factor ϕ_s and the Sauter mean particle diameter d_p will be different. Thus, C, will have different values either side of a coke-ferrous interface and consequently equation (5) should be viewed as the governing equation in each layer rather than the governing equation throughout the whole furnace. If we now make the additional assumption that the temperature in each layer is approximately constant, we obtain for the ℓth layer

$$\hat{C}_\ell \left[\frac{\partial}{\partial r}\left(\frac{1}{r}\frac{\partial\psi}{\partial r}\right) + \frac{1}{r}\frac{\partial^2\psi}{\partial z^2}\right] + \frac{\partial}{\partial r}\left[\sqrt{\left(\frac{\partial\psi}{\partial r}\right)^2 + \left(\frac{\partial\psi}{\partial z}\right)^2} \cdot \frac{1}{r^2}\frac{\partial\psi}{\partial r}\right]$$

$$+ \frac{\partial}{\partial z}\left[\sqrt{\left(\frac{\partial\psi}{\partial r}\right)^2 + \left(\frac{\partial\psi}{\partial z}\right)^2}\,\frac{1}{r^2}\frac{\partial\psi}{\partial z}\right] = 0 \tag{6}$$

where

$$\hat{C}_\ell = \frac{150\mu_\ell(1-\varepsilon_\ell)}{1.75\phi_{s,\ell}d_{p,\ell}}$$

Adjoining layers are coupled by the requirement that the pressure be continuous across the interface. Further boundary conditions must now be imposed. Due to symmetry, the radial velocity component is zero at the centre line

i.e. $\qquad \lim_{r\to 0}\frac{1}{r}\frac{\partial\psi}{\partial z} = 0$.

On the furnace wall and softened ferrous layers we require that the normal component of the velocity be zero,

i.e. $\qquad \frac{\partial\psi}{\partial s} = 0$

where $\underline{s}$ is the unit vector tangential to the boundary. In addition, the pressure at the stockline (the boundary of the uppermost layer) is constant. Finally, the mass flux will be specified at some portion of the boundary (at the coke slits for example).

Equation (6) with the appropriate boundary conditions was solved numerically, for the situation depicted in Figure 1 when the flow through the coke slits was specified. The geometry of each layer was transformed to a rectangle and the transformed nonlinear partial differential equations were then approximated by finite difference schemes. The boundary conditions were also discretized and an approximate solution to the resulting set of nonlinear equations was then obtained using a block successive over relaxation iteration. Some typical streamlines are shown on Figure 4 for the bottom four layers of an approximation to Hirohata number 1 furnace when the coke slit velocities had been calculated using another mathematical model (Jenkins and Burgess [2]). The most striking feature of such streamlines is that, away from the coke slits the flow becomes almost vertical. Given the degree of approximation involved in the model, (compare for example the idealized geometry in Figure 1 with actual geometries in Figure 2) it therefore seems reasonable to make the simplifying assumption that the flow in the coke slits is horizontal while the flow outside is vertical. However, this approximation should not be expected to be valid in regions where the layers are not approximately horizontal and of uniform width. In addition the effects of temperature gradients as can be expected in a furnace have not been considered.

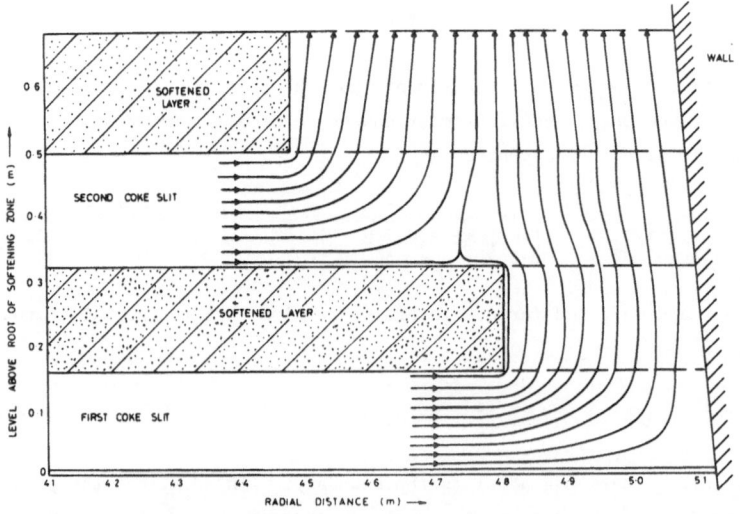

FIGURE 4 :   Predicted isothermal gas flow pattern for the bottom two coke slits in
             Hirohata No.1 furnace.

In the next section, we make use of the approximate result in a model that predicts the overall furnace pressure drop.

OVERALL FURNACE PRESSURE DROP MODEL

A model has been developed which predicts the distribution of the flow of gas through the coke slits of the cohesive zone in the blast furnace as well as the furnace pressure losses. This model assumes an idealised form for cohesive zone shape. Prediction of the gas flow distribution through the coke slits and the furnace pressure profile has been achieved by solution of the appropriate equations for gas flow through packed beds. The model is currently in the process of refinement in order to accurately predict blast furnace pressure profiles, and in the context of the present paper the model is a preliminary version.

Several assumptions have been made in the formulation of the model. These are necessary because technical details on softening and melting phenomena and solids flow mechanisms are not yet available for the blast furnace process.

The assumptions made are as follows :

1. The ferrous and coke layers are assumed to be horizontal throughout the furnace and the thickness of each layer is uniform across the furnace radius.

2. The size, voidage and bulk density of the coke and sinter particles are uniform, except for coke particles below the base of the cohesive zone where the size and voidage are uniform and lower.

3. The reduction in volume upon softening and melting is assumed equal to the void space of the unsoftened ferrous burden material and the softened material is therefore impervious to gas flow.

4. The assumed temperature distribution in the furnace is a linear function of depth in the furnace.

5. The gas flow direction is assumed vertical, except in the coke slits where a horizontal flow is assumed.

6. The volume flow rate of gas is assumed to be equal to the top gas volume in the region of the furnace above the outer boundary of the cohesive zone.

   The volume flow rate of gas in the bosh and cohesive zone regions equals the combustion gas volume. These gas flows can be computed for a particular furnace operation from a mass balance calculation.

7. The cohesive zone has been defined by specifying the positions of its inner and outer boundaries. The shape of these two curves has been assumed to be of the form

$$z = A + B \exp\left(- \frac{r^2}{2\sigma^2}\right) \tag{7}$$

where z is the depth of the curves from the stockline at radius r and σ is a
parameter.   Rearranging this equation gives the radial position of each curve as a
function of depth in the furnace as

$$r^2 = - 2\sigma^2 \ln \left(\frac{z-A}{B}\right) \qquad (8)$$

The constants A and B for each curve are determined by the following conditions :
for the inner curve (marked (1) in Figure 5)

$$r = 0 \quad at \quad z = H_1 + t$$

$$f = R_f(H_2) - W_B \quad at \quad z = H_2$$

and for the outer curve (marked (2) in Figure 5)

$$r = 0 \quad at \quad z = H_1$$

$$r = R_f(H_2) \quad at \quad z = H_2$$

where t is the thickness of the first softened ferrous layer, W_B is the width of
the bottom coke slit and $R_f(z)$ is radius of the furnace wall at depth z . The shape
and position of the cohesive zone is then totally defined by specifying the values of
H_1 , H_2 , W_B and σ .

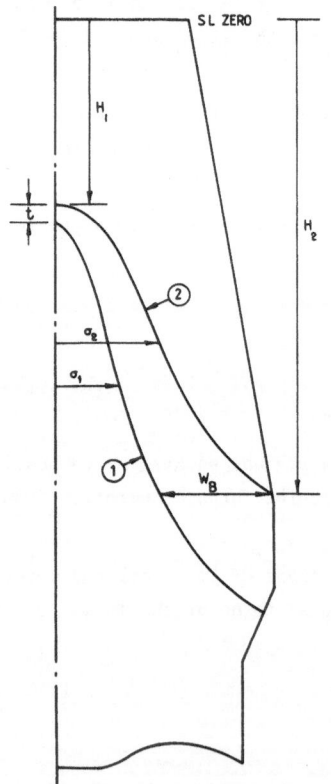

FIGURE 5 :   Cohesive zone
                    geometry.

8.   At any horizontal level in the softening-melting zone, some of the ferrous material has melted and disappeared, some has softened and a proportion remains as particulate material.   It has been assumed that coke flows into the void space created by the softening and compressing ferrous materials.   The layer thicknesses under these conditions have been obtained by calculating the remaining volumes of coke and ferrous material and dividing by the furnace cross-section.

9.   Pressure drops in the furnace zones are given by the following relations :

tuyeres pressure drop :

$$\Delta P_t = 0.75 \rho_t v_t^2 \tag{9}$$

bosh region pressure drop :

$$- \frac{dp}{dz} = 150 \frac{(1-\varepsilon)^2}{\varepsilon^3} \mu \frac{1}{(\phi_s d_p)^2} \frac{RT}{pM} G + 1.75 \left(\frac{1-\varepsilon}{\varepsilon^3}\right) \frac{1}{\phi_s d_p} \frac{RT}{pM} G^2$$

pressure drop in elemental packed beds in cohesive and lumpy zones :

$$\frac{P_1^2 - P_2^2}{L} = 3.5 \left(\frac{1-\varepsilon}{\varepsilon^3}\right) \frac{1}{\phi_s d_p} \frac{RT}{M} G^2 \tag{11}$$

Flow areas for the pressure drop calculation have been obtained as follows :

for vertically flowing gas :

$$A = \pi(r_2^2 - r_1^2) \tag{12}$$

for horizontally flowing gas in the coke slits :

$$A = 2\pi h \sqrt{r_1 r_2} \tag{13}$$

where r_1 is the inner radius of the cohesive zone coke slit, r_2 is the outer radius of the coke slit and h is the coke slit thickness.   The pressure drop calculation procedure for the cohesive zone requires solution of a set of simultaneous nonlinear equations describing the coke slit pressure losses, as in equations (11) and (13). An iterative numerical method has been used for this purpose.   The technique used is summarised in Figure 6.

Figure 7 shows a typical predicted result from the model showing the shape of the cohesive zone and the computed gas flow distribution through the coke slits as well as the furnace shell pressure distribution.   The assumed shape of the zone in this case is similar to that in Hirohata No.1 blast furnace (see Figure 2) and coke slit dimensions have been obtained by a volume balance at each level in the cohesive zone.   The

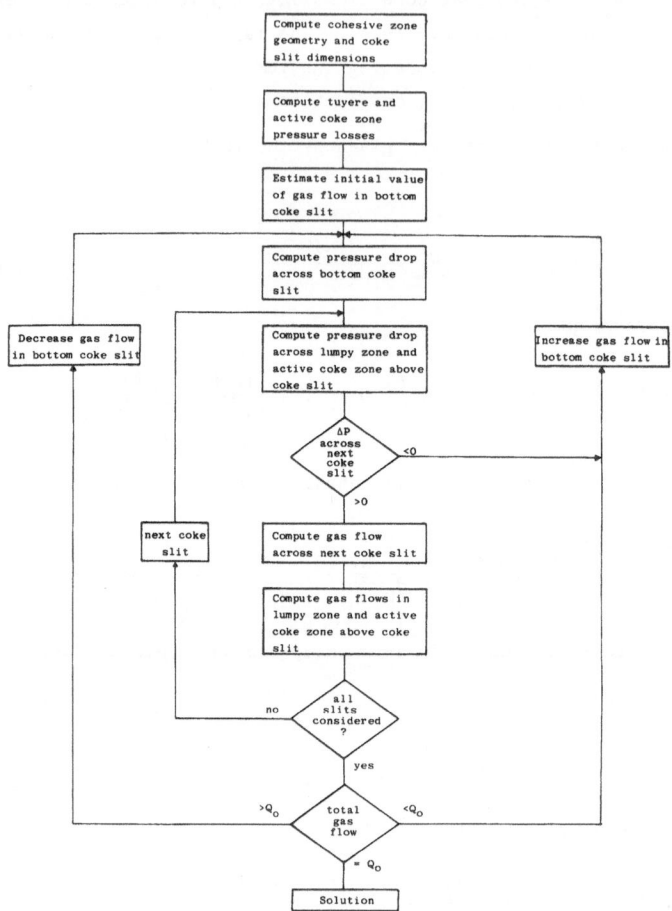

FIGURE 6 :   Schematic of solution procedure for cohesive zone coke slit gas flow
              distribution.

gas flow distribution through the coke slits exhibits a similar shape to that predicted
for Hirohata No.1 furnace by Nakamura *et al.* [2], with the greater proportion of gas
flow at the base of the cohesive zone.

The present model is finding application in the analysis of blast furnace gas flow
using measured pressure tappings on operating furnaces.   The pressure profile is
sensitive to the shape of the cohesive zone and work is now proceeding to predict the
way in which the furnace pressure distribution changes with change in the important
zone shape parameters.

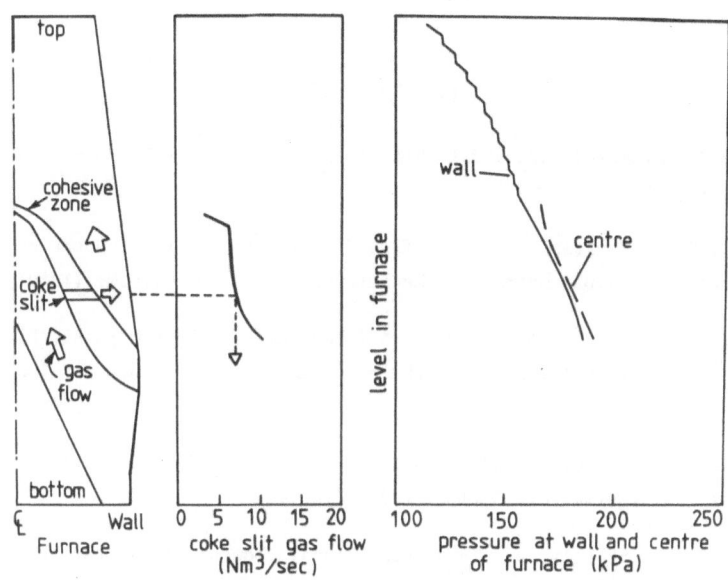

FIGURE 7 : Example of predicted cohesive zone gas flow distribution and furnace pressure profile.

CONCLUSIONS

The flow of gases within the ironmaking blast furnace has been modelled through the use of the Ergun equation, which describes the pressure loss for the flow of gas in a packed bed. In the first case, the vectorial form of the Ergun equation has been used to obtain a non-linear 2nd order elliptic p.d.e. for the stream function, with boundary conditions corresponding to flow out of the coke slits of the cohesive zone and upwards through the lumpy zone of the blast furnace. The results of the numerical solution of this problem have led to the formulation of a simpler model of gas flow through the cohesive zone. This latter model is based on the assumptions that gas flows horizontally within the coke slits and vertically in other regions of the blast furnace. Consequently, the problem reduces to the flow of gas through a series of interconnected packed beds of varying geometry. The solution to the problem then requires the simultaneous solution of a set of nonlinear equations for pressures and flow rates within the cohesive zone.

The results have indicated that predictive models to describe the distribution of gas flow in the blast furnace are useful for the analysis of the fundamentals of the process. Application of the models to a furnace pressure loss simulation, however, requires accurate data on coke and ferrous material particle sizes and void fractions within the furnace.

Further development of the models described will include application of the interconnected packed bed simulation to an operating blast furnace to accurately predict the
observed pressure profile and development of the vectorial Ergun equation to include non-
isothermal and heterogeneous packed beds.

REFERENCES

[1]   Bird, R.B., Stewart, W.E. and Lightfoot, E.N. Transport Phenomena, John Wiley & Son,
       (1960), 196-200.

[2]   Jenkins, D.R., Burgess, J.M. Gas flow distribution in the ironmaking blast furnace,
       Proc. Australia-Japan Extractive Metallurgy Symp., Aus. I.M.M. (1980), 257-268.

[3]   Kanbara, K., Hagiware, T., Shigemi, A., Karayama, Y., Wakabayashi, K. and Hiramoto, N.
       Dissection of blast furnaces and their internal state, Trans. I.S.I.J., 17 (1977),
       371-381.

[4]   Nakamura, N., Togino, Y. and Tateoka, M. Behaviour of coke in large blast furnaces,
       Ironmaking & Steelmaking, 5 (1), (1978), 1-17.

[5]   Stanek, V. and Szekely, J. Three-dimensional flow of fluids through nonuniform packed
       beds, A.I.Ch.E. Journal, 20 (5), (1974), 974-980.

LIST OF SYMBOLS

A	flow cross-sectional area	m^2
c_1, c_2	coefficients in Ergun equation (see equation (2))	-
d_p	Sauter mean particle diameter	m
G	Gas mass flow per unit area	$kg/m^2 s$
h	coke slit thickness	m
H_1	depth of top of cohesive zone	m
H_2	depth of bottom coke slit of cohesive zone	m
ℓ	index of layer	-
L	flow length	m
M	molecular weight	-
P	gas pressure	Pa
Q	flow rate of gas	m^3/s
r	radial co-ordinate	m
R	gas law constant	$m^2/s^2 k$
t	softened layer thickness	m
T	temperature of gas	K
v	superficial velocity of gas	m/s
$|\underline{v}|$	magnitude of velocity	m/s
W_B	width of bottom coke slit	m
z	axial co-ordinate	m
ε	voidage	-
μ	viscosity of gas	Pa.s
ρ	density of gas	kg/m^3
σ	parameter for defining cohesive zone shape (equation (7))	m
ϕ_s	particle shape factor	-
ψ	stream function	kg/s

THE MODELLING OF STRUCTURAL FRAMES SUBJECTED
TO DYNAMIC LOADING: A PARTICULAR APPLICATION

P. Swannell,
Divil Engineering, University of Queensland,

AND
C.H. Tranberg,
Macdonald, Wagner and Priddle, Pty Ltd, Brisbane.

SUMMARY

This paper describes the use of a mathematical model to predict the dynamic response of the new Gateway Bridge across the Brisbane River. The bridge is currently being designed and the procedures described below are concerned with the performance of the bridge under impact from a ship collision. Following the description of the bridge, reasons are given why prediction of the dynamic response would be virtually impossible without the use of a carefully chosen mathematical model. The paper concludes with a brief description of the relevant mathematics and the results obtained from the analysis.

1. INTRODUCTION

The structural engineer is called upon to design engineering structures which, safely and economically, will resist loads placed upon them from all sources which are likely to interact with them during their service lives. The engineer must also understand and make provision for possible but unlikely events which, if they occurred, might otherwise have catastrophic effect upon structural integrity. Rarely is it possible to achieve a satisfactory design solely by inspiration or good fortune. A satisfactory design will only emerge after a rational investigation based upon sound engineering science moderated where necessary by the experience of the engineer. The language and the basis of rational engineering science is sound mathematics and the designer will rarely be able to verify the appropriateness of intuitive schemes without the application of mathematical techniques of analysis. Even in the most simple structures there is consolation to be had from rational check analyses of any intuitive solution.

Most engineering structures receive their loads from natural sources such as wind, waves, earthquakes etc, from imposed service loads such as vehicles, occupancy, machinery etc, or from undesirable but possible accidental external agencies such as impact from moving bodies.   Almost without exception the applied loads are *time dependent* and produce structural responses which are also *time dependent*.   Factors which influence response of the structure include the mass and stiffness of the structure and the nature of the time dependency of the imposed loads themselves.   In this dynamic environment intuitive understanding of the structural response is extremely difficult if not impossible to achieve.   It would be exceedingly difficult to describe the likely response of a structure without resorting to mathematics.   It is also worth noting that it is extremely difficult to model *dynamic* response in a physical way using other than a full-size structure.   Physical "modelling" by use of a special-purpose prototype, as with an aircraft for example, is rarely conceivable in civil engineering structures.

Mathematics, then, is an integral part of the structural design process and mathematical modelling, in a dynamic context, can be divided into three aspects.   The *first* is the description of the *input* to the system, i.e. the time dependent loads or imposed displacements.   The *last* is the *output* from the system, i.e. time dependent structure displacement and stresses. The manner in which output varies with time is generally different from the manner in which input varies.   Reference [6] makes this aspect of the problem clear in terms of relevant mathematics.

A major factor in arriving at output response which is both useful and, in any physical sense, 'correct' is the second aspect of an analysis procedure.   Input is "transformed" into output by use of a suitable mathematical model of the real structural system.   Many different models may be used and each will produce output which need not necessarily have either theoretical or physical merit.   The study of suitable mathematical models of structural behaviour is an art in itself and calls for a combination of sound mathematical skill and deep understanding of the physical significance of the various parameters that might be necessary in an adequate description.

The Gateway Bridge project is an exciting new project currently under construction in Brisbane and provides an excellent example of a structure whose final form and adequacy can only be determined by the careful use of appropriate mathematics.   The project is described very briefly below and this description is followed by an account of the analysis of the bridge under the impact of a ship collision with the bridge at pile cap level producing time dependent in-plane response of the bridge.

2. THE GATEWAY BRIDGE

2.1 Project Description

The Gateway Bridge now under construction in Brisbane is a vehicular crossing of the lower reach of the Brisbane River between Queensport and Eagle Farm.   Design of the bridge is being undertaken by Macdonald, Wagner and Priddle Pty. Ltd. in association with Losinger Ltd.

of Berne, Switzerland.   Construction is by Transfield (Qld) Pty. Ltd. and the project is due
for completion in early 1985.   The complete river crossing consists of a prestressed concrete
bridge 1627 m between abutments with associated approach road works linking the bridge to
existing roadways.   It will carry six lanes of traffic with five approach spans on the
southern bank, three river spans and a further ten approach spans on the northern bank.
Figure 1 shows a general arrangement drawing of the structure.   The main river span of 260 m
will be the longest span for this type of bridge yet constructed.

The river spans comprise a continuous frame between expansion joints linking them to
the approaches.   A single cell box is employed for this portion of the bridge superstructure,
which varies in depth from 15 m at the river piers to 5.2 m at midspan.   The box superstructure is supported at each river pier by twin concrete columns which rise from submerged
pile caps, providing an elegant and aesthetically pleasing profile.

River piers are founded on piled footings each consisting of some fifty cast-in-situ
reinforced concrete piles, taken down into sandstone bedrock.   Considerable attention has been
paid to protection of piers and pilecap from damage by ship impact normal to the plane of the
bridge.

The feature of the bridge is the requirement that it must fit within a very shallow
"window" defining sufficient navigation clearance for shipping passing below the bridge and a
maximum height permissible with regard to air traffic requirements.   At midspan the main deck
is elevated 64.52 m above datum and the maximum height of the bridge superstructure is limited
to less than 80 m above datum.

The extensive approach spans on both river banks consist of twin cell prestressed
concrete box superstructure of similar exterior profile to that of the main river span.   The
approaches are to be constructed at a 5.3% grade, and consist of fully continuous spans
generally of 71 m.   The box superstructure depth is 3 m in general, but is increased to 5.2 m
over the final spans to match the main cantilever box depth at the expansion joints.

Reinforced concrete box columns hinged top and bottom will support the superstructure.
The columns are founded on high level spread footings on the southern bank, and on prestressed
concrete piled footings on the northern bank.

2.2 Design for Ship Impact in the Plane of the Bridge

A design feature of the Gateway Bridge is the protection afforded to the piers and
pilecaps.   Nevertheless a design requirement is that the structure shall be capable of
resisting a horizontal impact of 20,000 kN applied at pile cap level in a span-wise direction.
This impactive force is time-dependent and will be resisted partially by the deflexion of the
pile group and partially by the inertia and structural integrity of the superstructure.   Under
the impact the structure will respond in a time-dependent manner and can be modelled as an
elastic system constrained by springs of known stiffness.

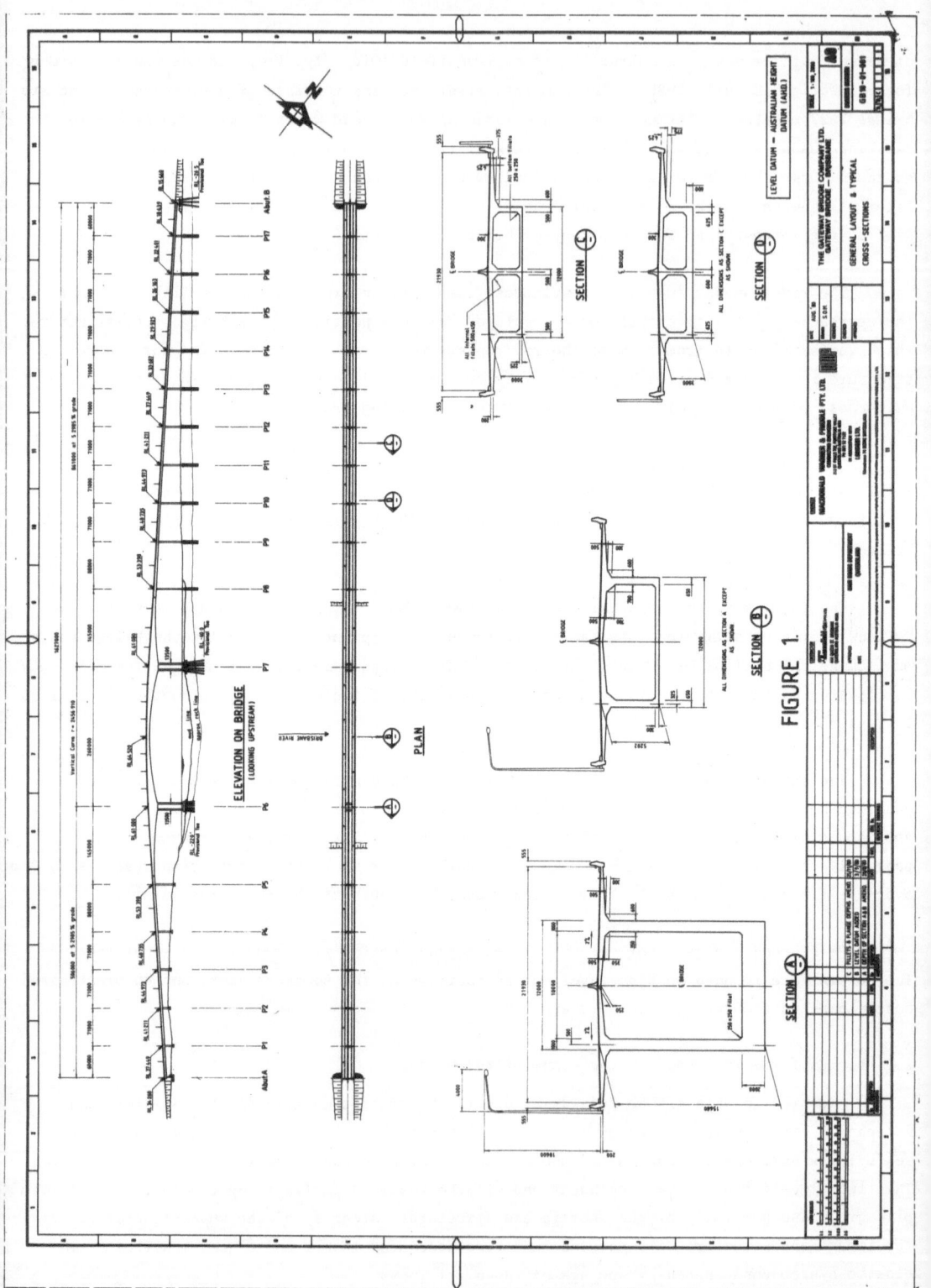

FIGURE 1.

Suitable spring constants for the representation of support stiffness were derived by conventional analysis of the pile groups. The modelling of the time-dependency of the impactive force due to a ship collision is a complex matter. The idealised forcing function eventually chosen is shown in Figure 2. References [4] and [5] give guidance in this area.

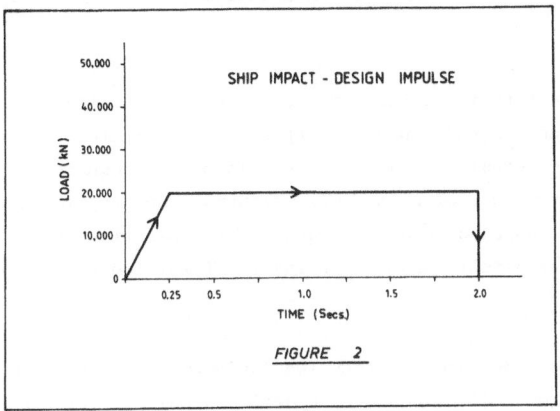

FIGURE 2

With a knowledge, then, of the proposed bridge geometry, a description of idealised support conditions, and a synthesised accident condition, the problem is specified. It is not inappropriate to observe at this stage that recent trends in "fringe-technical" education emphasize the need to specify problems. They also tend to ignore the need to learn sufficient technical skills for safe arrival at successful solutions to those problems. It is the engineer's task, almost uniquely, not only to specify a problem but also to proceed to a satisfactory solution ...... and suitable mathematics are now unavoidable.

A further observation is also valuable. In specifying the problem all aspects *cannot* be defined with equal accuracy or with equal confidence. This is always the case when modelling physical systems. It is the engineer's responsibility to assess the quality of the data and thence to judge the reliability of the conclusions derived from the data. The current problem is a good example.

With some reservation about material properties, the "reliable data" are the facts of the bridge geometry. The "less reliable data" are the facts of the support stiffness (soil mechanics is a complex discipline). The "least reliable data" are the facts of the impacting load. The engineer must view the problem, then, not simply as an application for mathematics, but as the description *with inadequacies* of a physical system. The use of good mathematics and sound computational procedures assist in making the theoretical model most nearly possess the true characteristics of the real system, despite such inadequacies.

It follows that it would be quite erroneous to argue that a simple unsophisticated model of the system would be "good enough" because of the variable quality of the data. Poor data certainly cannot be transformed into superb output by use of a sophisticated system model. However engineers do not, essentially, work with "poor data". They work with the "best available data" and the best available data deserves the best possible treatment. The real danger is that the best available data can be transformed into the least valuable output by the use of inadequate, insensitive or grossly simplified system models.

3. THE CHOICE OF A SUITABLE MODEL

3.1 Options Available

The problem under investigation calls for the choice of a suitable description of the structural system such that an assumed input forcing function (Figure 2) is transformed into a reliable description of the actual response. The information required for design purposes is a set of time-dependent structure displacements, structure bending moments, axial loads and shearing forces. These response data are required at close time intervals immediately following the impact. The transient response during the first two seconds is unlikely to be affected by any structural damping.

In choosing a suitable model and strategy two fundamental decisions must be made, in the context of an elastic, "small displacement", dynamics problem. Firstly, a choice must be made regarding the type of element which will form the basis of the mathematical model. Thereafter a choice must be made between alternative solution strategies, viz., use of a modal superposition technique describing response by weighted mode shape contributions or, alternatively, solution of the discretised governing differential equations of motion in terms of the actual nodal degrees of freedom.

In very broad terms, the analyst is required to assess the shortcomings of any strategy. Numerical methods for the solution of the governing equations are well-tried. Direct integration schemes, from the pioneering work of Newmark onwards are well documented and very valuable. More recently (Reference [7]) Finite Integral methods have also been used with some success. Broadly speaking the ease with which "answers" may be obtained depends upon the length of the time-step adopted in the discretisation. This is in turn influenced by the dynamic characteristics of the time-dependent loads and the structure itself.

The use of modal super-position techniques in the context of lumped-mass idealisations is common in relatively small-scale problems. It is less well-known in the context of the more sophisticated "exact" element described by the present authors in References [6] and [8]. All modal superposition techniques suffer from the difficulty that the analyst must use a "truncated set" of modal shapes, assessing "how many" modes should be incorporated in the solution.

The free vibration analysis which must precede any forced vibration analysis when a modal super-position is adopted may be of two kinds, determined by the type of element chosen as the "basic building block".

If the beam element in a frame analysis is constrained to vibrate, for example, such that *spatially* its displacements vary in a *prescribed* way irrespective of the time variation, then the resulting stiffness and mass matrices "separate out" into an elastic stiffness matrix

and a *consistent mass* matrix.   The free vibration description is said to be *linearised* in the
sense that the force contributions due to accelerations of the mass of the structure are found
to be proportional to the square of the vibrating frequency.

Similarly, in a simple lumped mass idealisation in which the structure is modelled as
a series of massless springs inter-connecting "point lumps", the force contributions due to
acceleration are again proportional to the square of the vibrating frequency.

In each of the above cases the equation describing the free vibration is a polynomial
in (frequency)^2 with solutions equal in number to the number of degrees of freedom of the
system, and a similar number of (very approximate) mode shape predictions.   Further, accuracy
of the predictions of vibrational behaviour depends upon the number of elements in the
idealisation.   In brief, the number of modes, irrespective of their "correctness", available
to the analyst is limited to the number of nodal degrees of freedom in the structural
idealisation.

The present authors have detailed in Reference [6,8] *et al.* "exact" dynamic stiffness
matrices describing the free vibrational behaviour of beam-columns, by solving the differential
equations of motion without need to constrain the element with regard to its internal
displacement distribution.   The resulting structure stiffness matrix contains coefficients
which are transcendental functions of the vibrating frequency.   The value of the determinant
of the structure stiffness matrix becomes infinite at particular (and fascinatingly significant)
values of the vibrating frequency.   A plot of the determinant value versus the eigen parameter
is discontinuous via $\mp$ infinity.   There are an infinite number of solutions to the
characteristic equations and (as one would properly expect) an infinite number of natural modes.
The analyst, potentially, has an infinite number of modes available to him from which to select
the truncated set.   The predictions of free-vibrational response are unaffected by the number
of *elements* chosen in the idealisation since the infinite-degree-of-freedom system is properly
modelled.

It will be recognised that, whereas lumped mass and consistent mass models lead to a
'linear' or 'algebraic' eigenvalue problem, the transcendental formulation gives rise to a
'non-linear' eigenvalue problem, with its inherent computational challenges!   Wittrick and
Williams (Reference [9]) provide the key to the satisfactory solution of such problems.

The results described in Section 4 are achieved by use of a modal super-position
strategy in the context of the "exact" distribution parameter (mass and stiffness) Timoshenko-
Rayleigh-Euler beam column, i.e. a beam-column model which includes the destabilising effect of
static axial load (Euler), shearing deformation (Timoshenko), rotatory inertia, lateral and
axial vibration (Rayleigh).

3.2 Solution Procedure

The methodology and appropriate mathematics is, briefly, as below :

(i)   The actual structure is modelled as a series of prismatic beam-column elements inter-connected at nodes and spring-supported.

(ii)   Element stiffness matrices modelling freely vibrating elements possessing distributed mass and stiffness are assembled into an overall structure stiffness matrix and constrained.

(iii) Solutions to the non-linear eigenvalue problem, incorporating the frequency, ω, of free vibration as the eigenparameter, are established for as many natural frequencies as are necessary for an adequate solution.   An infinite number of solutions is potentially available!

(iv)   Natural modes are computed and the time-dependent weighted contributions of each mode to both the total displacements and the inertia displacements of the structure are evaluated for a given forcing function.   This is achieved by the formulation and solution of an equivalent single degree of freedom system appropriate to the particular applied forcing function after the manner of Hurty and Rubenstein (Reference [2]).

(v)   Final structure bending moments, shearing forces and axial loads are computed as the sum of the "massless" (static) response and the modifying effects due to inertia.

With the aid of Wittrick's algorithms the computational procedures are reasonably easy to establish.   Without them, a "fool-proof" free vibration analysis is not achievable.   The idealisation is "exact" within the meaning described to that word in classical, elastic, small displacement analyses.   The solution accuracy is independent of the number of elements used in the model.   All aspects of the procedure are "analytic" without recourse to numerical approximations at any stage.

The essential procedure can be described by four governing equations with associated subsidiary definitions.   They are summarised below and fully detailed in Reference [6].

$$r(t) = {}_{I}r.g(t) + {}_{II}r(t) \tag{1}$$

$$_{I}r.g(t) = K_{s}^{-1}\bar{R}.g(t) \tag{2}$$

$$_{II}r(t) = \sum_{i=1}^{\infty} {}_{II}q_{i}(t).A_{i} \tag{3}$$

$$s(t) = k_{s}.a.{}_{I}r.g(t) + \sum_{i=1}^{\infty} {}_{i}k_{D}.a.{}_{II}q_{i}(t).A_{i} \tag{4}$$

where $r(t)$ = Final structure nodal displacements

$_{I}r.g(t)$ = "Massless Structure" displacements

K_{s} = 1st Order Non-Linear Elastic Structure Stiffness Matrix

$\bar{R}$ = Amplitudes of applied nodal loads

$g(t)$ = Time Variation of applied nodal loads

$_{II}r(t)$ = Displacements due to inertia effects

A_i = ith Eigenmode

$_{II}q_i(t)$ = Time-dependent "weighting" of the ith eigenmode contribution to inertia effects

$s(t)$ = Final element end forces

k_s = Super-diagonal array of non-linear elastic static member stiffness matrices

$_ik_D$ = Super-diagonal array of the ith mode generalised dynamic member stiffness matrices

a = Structure topological matrix

$_{II}q_i(t)$ is given by the equation

$$_{II}q_i(t) = q_i(t) - {}_Iq_i(t) \tag{5}$$

where

$$_Iq_i(t) = \frac{\bar{F}_i(t)}{\bar{K}_i} \tag{6}$$

and $q_i(t)$ is given as the solution of the equation

$$\bar{M}_i\ddot{q}_i(t) + \bar{C}_i\dot{q}_i(t) + \bar{K}_iq_i(t) = \bar{F}(t) \tag{7}$$

$\bar{M}_i$, $\bar{C}_i$, $\bar{K}_i$ and $\bar{F}_i$ are the Modal Mass, Damping, Stiffness and Force in the ith mode respectively.

These equations have precise significance in an understanding of the interface between an "elegant piece of mathematics" and a "physically helpful description".

Equation (1), (2) and (3) properly describe the problem as the super-position of "massless" response (which can be obtained by a single "one-off" static analysis using the static stiffness K_s and equation (2) and the effects of inertia forces. Equation (3) properly describes the weighting of these inertia force contributions using the time-dependent weighting function $_{II}q_i(t)$ and the mode shapes A_i which have been derived from the non-linear eigenvalue problem. Equation (4) formally describes the transformation of displacements into member force effects and Equations (5) and (6) complete the solution procedure by defining the weighting function $_{II}q_i(t)$ and emphasising that they are *part* of the generalised coordinates defined by equation (7). Equation (7) is the one equation with which all dynamicists are familiar being the description of an equivalent single-degree-of-freedom system.

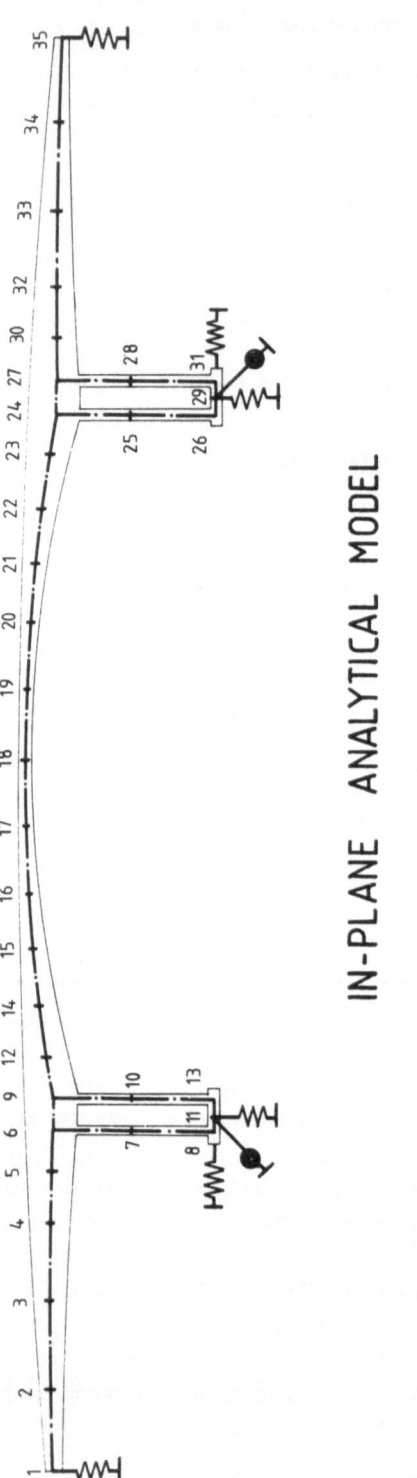

IN-PLANE ANALYTICAL MODEL

FIGURE 3

GATEWAY BRIDGE - IN-PLANE VIBRATION MODEL

NATURAL FREQUENCY (HZ)

Mode No.	Frequency	Mode No.	Frequency
1	0.1772	26	9.3970
2	0.5984	27	9.4716
3	1.1066	28	10.4912
4	1.3988	29	11.0662
5	1.7750	30	11.3211
6	2.3267	31	12.4588
7	2.4122	32	13.1707
8	2.4169	33	13.2608
9	2.4417	34	13.4458
10	2.6997	35	14.1017
11	3.2266	36	14.4452
12	3.3090	37	14.9596
13	3.4862	38	15.3497
14	4.2959	39	16.2699
15	5.2891	40	17.3734
16	5.4818	41	18.0510
17	5.9155	42	18.0526
18	6.0616	43	18.6634
19	6.2714	44	20.3358
20	6.7439	45	21.2699
21	6.8092	46	21.2916
22	7.3065	47	21.8263
23	8.2251	48	21.8496
24	8.8825	49	22.3221
25	9.2058	50	22.5754

FIGURE 4

4. SOLUTIONS

Figure 3 shows the idealisation of the structure. Nodes have been placed at points where significant changes in section properties occur and at actual structural joints. Tapering sections have been idealised with a suitable number of stepped prismatic sections. Nodes have been numbered so that the maximum difference between joint numbers at the ends of any one element is a minimum. This ensures that the band-width of the stiffness matrix will be as small as possible. Only terms of the stiffness matrix which lie on its leading diagonal and above it within the band-width are stored. All the results have been obtained by use of the authors' program FORVIB (Reference [3]) which uses a Rayleigh-Timoshenko-Euler exact beam-column as its basic element. Preliminary studies suggested that satisfactory forced vibration

results could only be achieved by use of a large number of mode contributions.   The reason
for this is that low natural frequencies are associated with modes which do not incorporate
significant movement of the spring bases whereas higher frequencies introduce this type of
motion.   Eventually it was found that the first *fifty* modes were sufficient to give results
which could be accepted with confidence.   It is, again, not possible to be totally certain
that there are no still higher modes which could contribute significantly but the authors'
judgement is that this is unlikely.

This particular investigation is an *excellent* example of the need to use sophisticated
models of the structure.   "Traditional" lumped mass models which, with relatively few lumps,
might be adequate for routine, basically rectangular, structures, could easily be misleading.
The prediction of Bending Moments and other force effects, in addition to deflexions, will
depend upon the detailed accuracy of the predicted contributions of higher modes to the shape
of the displaced structure.

Figure 4 gives the first fifty Natural Frequencies of the structure.   Figures 5 and 6
show some significant displacement results.   Figures 7-10 show some results for Bending
Moments at certain important positions on the bridge.   It is evident that the transient
response of the super-structure is sufficiently complex to make prediction by use of a simple
unsophisticated model an unlikely proposition.   Finally, Figures 11 and 12 give shear and
axial load responses at the base of the southern column.

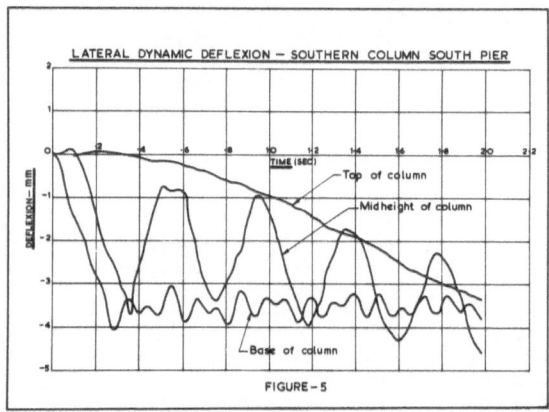

FIGURE 5

FIGURE 6.

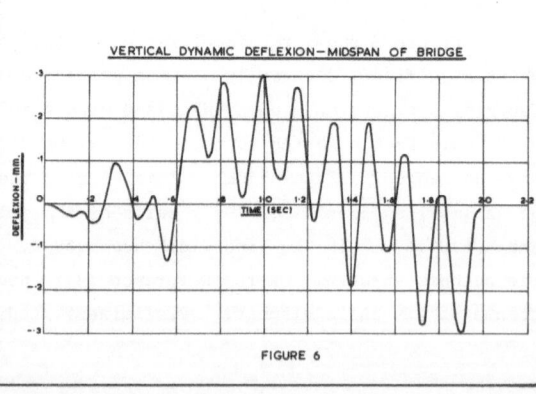

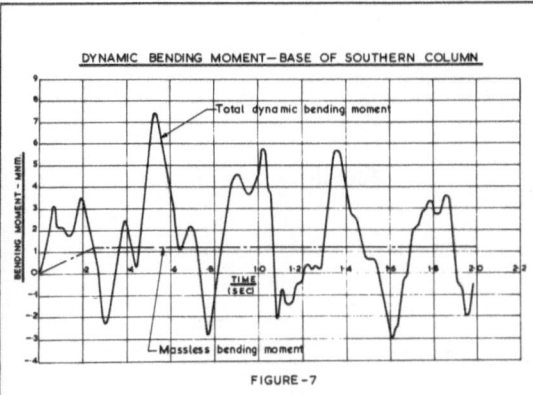

FIGURE 7

FIGURE 8

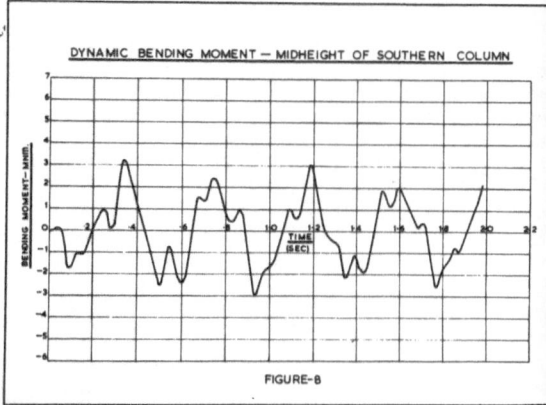

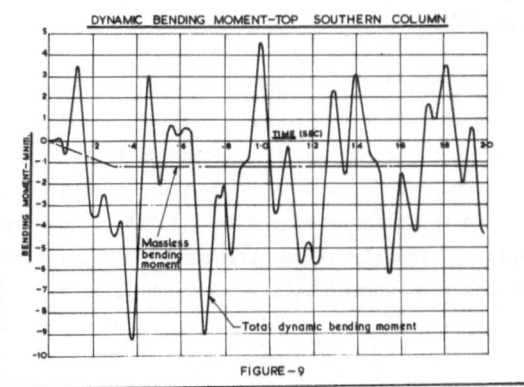

FIGURE 9

FIGURE 10

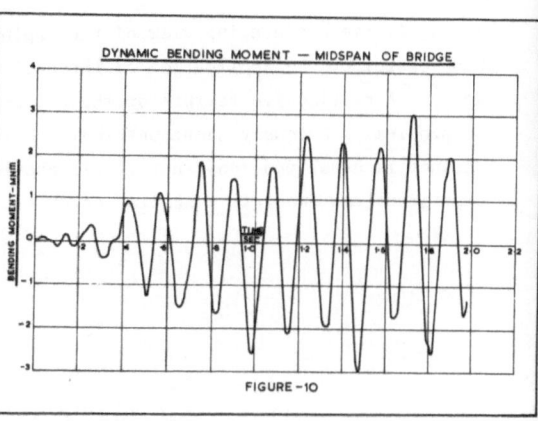

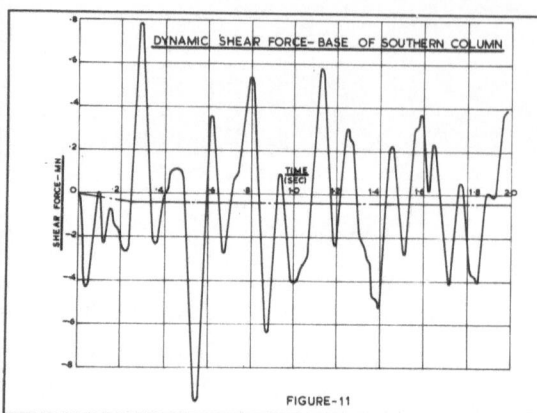

FIGURE 11

FIGURE 12

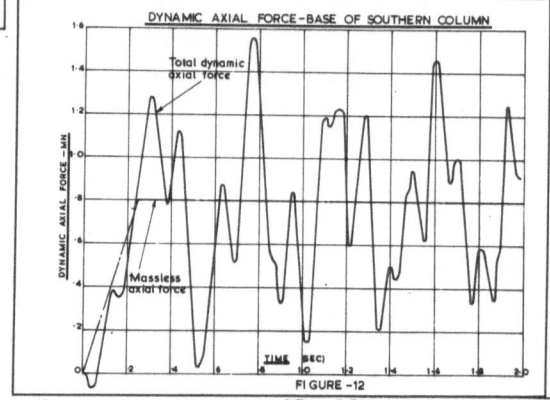

5. CONCLUSION

It has been argued in a recent paper (Reference [1]) that dynamic response can be predicted "as accurately" using simple empirical formulae as by "computer-based methods". Such an assertion if justified at all, could only be sustained in the context either of unrealistically simple structures or in extremely complex structures for which no theoretici would *claim* to have an adequate model. Indeed, in the data presented in [1], the structure under consideration contained aspects which could not readily be modelled and the theoretica results were quoted on the basis of theoretical analyses whose quality, through lack of information, could not be judged.

A more encouraging view of the engineer's task is that it is an intensely difficult o made easier by the application of the best available engineering science to the best availab data. A distinctive feature of the professional engineer's work is its demand for solution to problems previously unencountered. Mathematics, as the language through which engineeri science is developed, does not itself provide an engineering solution. When combined with experience and innovative skills, however . . . .

6. REFERENCES

[1]   Ellis, B.R. An Assessment of the Accuracy of Predicting the Fundamental Natural
        Frequencies of Buildings and the Implications concerning the Dynamic Analysis of
        Structures. *Proc. Inst. Civ. Engrs.* Part 2, 1980, 69, Sept. 763-776.

[2]   Hurty, W.C. and Rubenstein, M.F. *Dynamics of Structures.* Prentice-Hall, Englewood
        Cliffs, NJ, 1964.

[3]   Macdonald, Wagner & Priddle Pty. Ltd. FORVIB : A FORTRAN package for the Analysis of
        the Free or Forced Vibration of Frames. MWP Pty. Ltd., Brisbane, Qld 1978.

[4]   Ostenfeld, Chr. Ship Collisions Against Bridge Piers. *IABSE* Vol. 25, 1965 pp. 233-278.

[5]   Statsbroen Store Baelt Copenhagen. Investigation into the Ship Collision Problem :
        The Great Belt Bridge, February 1979.

[6]   Swannell, P. and Tranberg, C.H. Procedures for the Forced, Damped Vibration Analysis
        of Structural Frames using Distributed Parameter Models. *Computer Methods in App.
        Mech. and Engg.* 16 (1978) pp. 291-302. North-Holland Pubs. Co. Amsterdam.

[7]   Swannell, P. The Solution of Forced Vibration Problems by the Finite Integral Method.
        Univ. Qld. Civ. Eng. Dept Research Report No. CE16, August 1980.

[8]   Tranberg, C.H. On the Free and Forced Vibration of Structural Frames. Ph.D. Thesis,
        Dept Civ. Eng., Univ. Qld. 1977.

[9]   Wittrick, W.H. and Williams, F.W. A General Algorithm for Computing Natural
        Frequencies of Elastic Structures. *Mech. and App. Maths.* Vol. XXIV, Pt. 3,
        pp. 263-384, 1971.

BLAST FURNACE HEARTH DRAINAGE

W.V. PINCZEWSKI AND W.B.U. TANZIL,

Department of Chemical Engineering,
University of New South Wales,

AND

J.M. BURGESS AND M.J. McCARTHY,

Central Research Laboratories,
Broken Hill Proprietary Limited, Newcastle.

ABSTRACT

The effective drainage of liquids from the hearth is a major problem in the operation of large commercial blast furnaces. Because of the nature and complexity of the blast furnace process, mathematical modelling provides a powerful means of obtaining the insight into the drainage process which is necessary for this problem to be overcome.

A mathematical model which allows a prediction of the position of the gas-liquid interface during the drainage of a two-dimensional packed bed is described. The calculations are compared with drainage experiments conducted on a laboratory scale packed bed and the agreement between the two results is shown to be excellent. On the basis of the computations carried out it is concluded that hearth drainage is improved by the adoption of a continuous tapping procedure, an improvement in coke quality and a reduction in slag viscosity.

INTRODUCTION

Blast furnaces account for almost all of the world's primary steel production and at present there are some 1,000 furnaces operating worldwide producing some 500 million tonnes of molten iron per annum (Peacey and Davenport (1979)). The furnace itself is a tall vertical reactor in which iron ore is rapidly reduced to iron. A schematic view of a typical modern blast furnace is shown in Figure 1. For purposes of discussion it is convenient to subdivide the furnace into three major zones - the *shaft,* the *bosh* and the *hearth*. The relative locations of these zones are shown in Figure 1, together with a sub-zone called the *fusion* or

cohesion zone which extends from the bosh into the shaft of the furnace.   A detailed
description of these zones is given in a companion paper by Burgess *et al* published in these
proceedings.

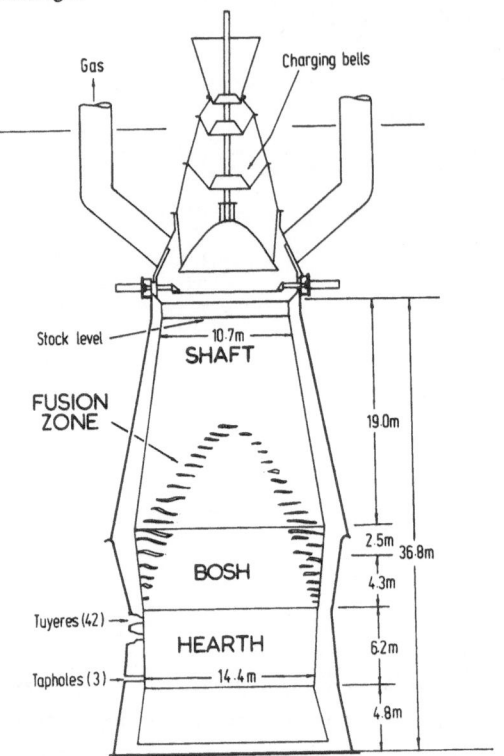

FIGURE 1 : Schematic view of a
typical modern blast furnace
(Fukuyama #5 N.K.K., 4617m^3).

Briefly, raw materials - iron ore (Fe$_2$O$_3$) , coke (C) and flux (CaO/MgO) - are
introduced in layers at the top of the furnace through a system of gas locks or bells and
descend through the shaft of the furnace.   Hot reducing gases rising from the bosh zone heat
the descending material and chemically reduce the ore to metallic iron.   *Hot blast* air, at
a temperature of 1200 - 1600°K and usually enriched with oxygen and/or hydrocarbon fuel, is
injected through *tuyeres* located around the circumference of the furnace at the top of the
hearth and reacts with incandescant coke to produce the carbon monoxide required for the
reduction reactions in the shaft zone.   This gasification reaction is highly exothermic and
results in temperatures in the range 1700 - 2400°K in the bosh and hearth zones.   Temperature
in the cohesive zone are sufficiently high for the descending mixture of reduced and partiall
carburized iron and slag forming fluxes to gradually soften and melt.   The molten iron and
slag trickle downwards and accumulate in the hearth from where they are periodically or
continuously drained out of the furnace through *tap holes* located near the bottom of the
hearth.

The cost of a typical modern blast furnace (5000 tonnes molten iron/day) is approxima
$150 million and the campaign life of such a furnace is of the order of 5 years.   After thi
time the furnace lining requires replacement and the cost of this operation is approximately

$60 million.   In addition to the high capital cost, operating costs have increased
dramatically over the past decade as a direct result of increases in the cost of raw materials
and energy.

Because of the high costs involved, furnace productivity and efficiency are very
important considerations in blast furnace operation.   Over recent years considerable effort
has been made to improve overall furnace performance and this has resulted in a trend to the
construction of larger furnaces.   In 1967, the world's largest furnace was Yawata Steel
Company's Sakai No.2 with an internal working volume of 2600 m^3 . (Strassburger (1969)).
In comparison with this Nippon Steel Corporation has recently commissioned furnaces with
working volumes in the 5000 m^3 class (Nakamura *et al* (1978)).   The design of these large
furnaces has been largely based on previous successful operating experience with smaller
furnaces.   However, current experience shows that the larger furnaces are subject to particular
problems not encountered with the smaller furnaces.   A major problem is the effective drainage
of liquids from the hearth of the furnace.   Poor hearth drainage (i.e. the inability to
completely drain the hearth) leads to unstable furnace operation and this is generally
accompanied by a marked loss in furnace productivity and efficiency.   In order to understand
how this comes about, we must look more closely at conditions within the hearth itself.

Conditions in the hearth zone

The molten iron and slag which accumulate in the hearth are immiscible and as a result
of gravity they segregate into two distinct layers with the lower layer being the heavier iron
phase (specific gravity approximately 6.8) and the upper layer being the lighter slag phase
(specific gravity approximately 2.8).   Unreacted coke, which is the only solid phase present
below the level of the cohesive zone; forms a column or bed of packed coke particles which
extends into the hearth.   In the hearth this coke bed sinks into the molten iron and slag
pool to a depth which is determined by the equilibrium between the downward force exerted on
the coke column from above and the buoyancy force exerted on it by the molten iron and slag.
In smaller blast furnaces (hearth dimaters 5-8 m) the downward force on the coke column is
generally smaller than in larger furnaces and the buoyancy force is therefore sufficient to
prevent the coke column from penetrating deeply into the hearth.   Under these conditions there
is little resistance to the flow of liquid from the hearth and drainage through the taphole
is effective.   In larger blast furnaces (hearth diameters 8-14 m) however, the downward-force
on the coke column is greater and the buoyancy force is insufficient to fully support the coke
column.   Under these circumstances the coke column penetrates to the bottom of the hearth and
the hearth liquids must therefore drain through a packed bed of coke.   This condition is
depicted schematically in Figure 2.   The packed coke bed introduces a significant resistance
to the flow of liquids from the hearth and this makes it difficult to effectively drain the
hearth.

At the beginning of the tapping operation (i.e. when the taphole is opened) the molten
iron flows out first.   Because of its relatively low viscosity 5-7 mPa.s. (depending on
temperature and composition) the pressure gradients in the direction of the flow are very small

and the iron-slag and gas-slag interfaces remain essentially flat allowing the iron phase to drain down to the level of the taphole.   When the slag phase begins to flow, this situation changes.   Molten slag has a viscosity in the range 100-500 mPa.s. (again depending on temperature and composition) and as a result the slag flow is associated with appreciable pressure gradients in the direction of flow and these are sufficient to produce a marked tilt in the gas-slag interface towards the taphole.   The tapping operation is terminated when the gas-slag interface first reaches the taphole (i.e. when gas begins to blow out from the furnace).   Because of the tilt in the interface this will generally occur prior to all of the slag being drained from the hearth.   The volume of slag which remains undrained at the end of the tapping operation is called residual slag and this has a detrimental effect on furnace performance.   For any given tapping strategy - single or multiple hole tapping, continuous or periodic tapping - the higher the residual slag volume the higher the required liquid level in the hearth necessary to maintain stable operation of the furnace at a given iron production rate.   Since the volumetric production rates of slag and iron are approximately the same, an increase in iron production rate leads to an increase in the operating liquid level in the hearth.

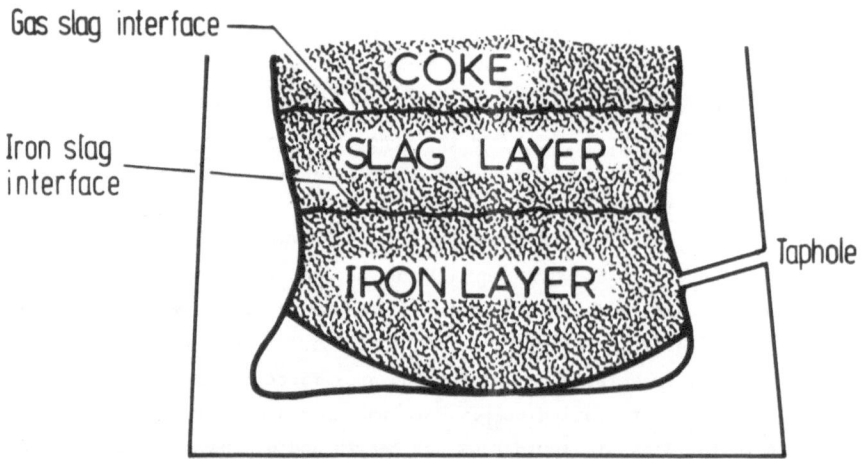

FIGURE 2 : Conditions in the hearth zone of a large blast furnace.

High liquid levels in the hearth are undesirable, since they cause major furnace operating problems.   If the liquid level approaches the level of the tuyeres the distribution of gas in the bosh zone is adversely affected and this reduces both furnace efficiency and furnace life.   Further, the levels of sulphur and silicon in the molten iron are closely related to the approach to chemical equilibrium of the slag and iron phases in the hearth.   A change in the relative residence times for the iron and slag phases in the hearth, which result from excessive slag holdup in the hearth, (high residual slag volumes), can adversely affect the chemical composition of the iron and lead to high costs in the downstream processing of the iron.   Finally, low liquid holdup in the hearth, i.e. low residual volumes and low residence times, assist in keeping the hearth hot and prevent local chilling and solidification of iron and slag. Local chilling and solidification reduce the area open to flow and result in high local velocities of iron and slag which can rapidly wear the refractory lining and hence reduce the campaign life of the furnace.

NEED FOR A MATHEMATICAL MODELLING APPROACH

To prevent the losses in furnace productivity and efficiency, which result from operation with high liquid levels in the hearth, we need to prevent excessive slag residual volumes. This requires a means of predicting residual slag volume as a function of both furnace operating conditions (production rate, tapping practice) and physical characteristics or quality of raw materials. Two general approaches are possible - an empirical or semi-empirical approach relying on actual measurements of operating liquid levels in a commercial blast furnace or a purely mathematical modelling approach based on a solution of the equations describing the physical processes which govern drainage behaviour in the hearth. The first approach has the apparent advantage of simplicity and would therefore appear attractive. Unfortunately the severe environment in the hearth makes it extremely difficult, if not physically impossible, to continuously minitor the liquid level in an operating commercial furnace. This leaves mathematical modelling as the only viable approach to predicting hearth drainage behaviour.

The hearth drainage problem is complex in that it requires the simultaneous solution of a system of strongly coupled nonlinear partial differential equations describing the processes of mass, momentum and energy transport for three fluid phases (iron, slag and gas) in three dimensions with moving boundaries (iron-slag and gas-slag interfaces). Although the problem is formidable, numerical techniques are available (Peaceman (1977)) which will, in principle at least, yield solutions to the problem.

As a preliminary to solving the full three-dimensional problem we have considered a simpler two-dimensional analogue which retains many of the important physical characteristics of the three-dimensional problem, but with a considerable reduction in computational complexity and time. The two-dimensional model allows us to identify the major physical and operational parameters which influence hearth drainage and results from this model may be compared with experimental results previously obtained by the authors (Burgess *et al* (1980)) for the drainage of two-dimensional packed beds. The experimental results were obtained as part of an ongoing investigation concerned with the characterization of the internal state of the ironmaking blast furnace.

The drainage system under consideration is shown in Figure 3. Since the highly viscous slag phase is almost entirely responsible for poor hearth drainage we need only consider a two-phase system (gas-slag). The packed bed may be non-homogeneous but it is assumed to be isotropic (i.e. intrinsic permeability may vary with position in the bed but at any particular position it is constant in all directions). The liquid is assumed to be incompressible and of constant density and viscosity.   The interface between the liquid and gas is assumed to be abrupt and capillary effects are neglected.   The pressure in the gas phase is assumed to be constant and the flow in the liquid is assumed to be viscous or laminar (i.e. inertial or turbulent dissipation is neglected).   Posed in this form the problem is similar to the free surface flow problems in porous media encountered in subsurface hydrology (Todson (1971)).

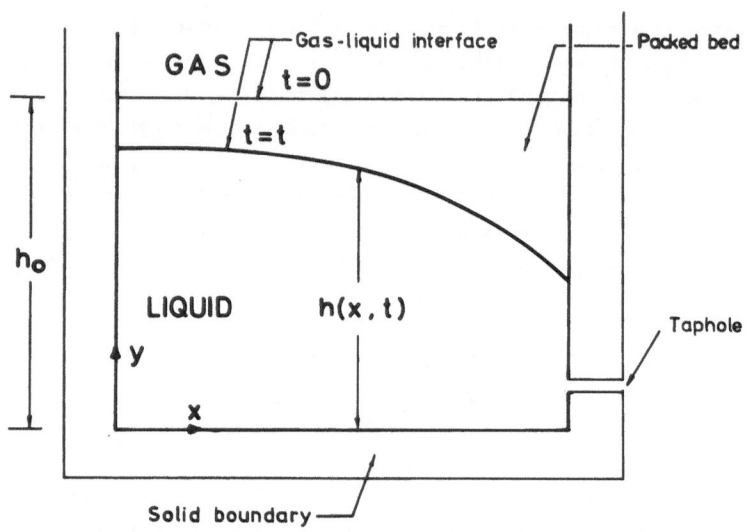

FIGURE 3: Simplified 2-dimensional drainage model.

The x and y components of the liquid superficial velocity are given by Darcy's equation as

x-direction

$$u = - K \frac{\partial \phi}{\partial x} \qquad (1)$$

y-direction

$$v = - K \frac{\partial \phi}{\partial y} \qquad (2)$$

where K is the hydraulic conductivity and ϕ is the hydraulic potential, defines as

$$\phi = \frac{P}{\rho g} + y \qquad (3)$$

where P is the pressure in liquid, ρ is the density of liquid, and g is the gravitaiona acceleration.

The equation of continuity may be written as

$$\frac{\partial u}{\partial x} + \frac{\partial v}{\partial y} + q = 0 \tag{4}$$

where q is the volumetric rate of outflow per unit volume of packed bed and is everywhere zero except at the drain point.

Combining Darcy's equation with the equation of continuity gives

$$\frac{\partial}{\partial x} \left(K \frac{\partial \phi}{\partial x} \right) + \frac{\partial}{\partial y} \left(K \frac{\partial \phi}{\partial y} \right) = q \tag{5}$$

which governs the pressure in the bed and which together with the appropriate boundary conditions may be solved to yield the position of the gas-liquid interface as a function of time.

The relevant initial and boundary conditions with reference to Figure 3 are

(i)   Initial condition

$$h(x, t_0) = h_0 \tag{6}$$

(ii)   No-flow boundaries

$$\frac{\partial \phi}{\partial x} = 0 \qquad \text{on AD and BC} \tag{7}$$

$$\frac{\partial \phi}{\partial y} = 0 \qquad \text{on CD} \tag{8}$$

(iii)   Free surface

$$P = 0 \, , \quad \phi = y \quad \text{on} \quad AB \tag{9}$$

The general form of the interface is given by

$$y = h(x, t) \tag{10}$$

where h is the surface co-ordinate.   The kinematic condition which must be satisfied at the interface is obtained by taking the total derivative of equation (10)

$$\frac{dy}{dt} = \frac{\partial h}{\partial x} \frac{dx}{dt} + \frac{\partial h}{\partial t} \tag{11}$$

Since $\frac{dx}{dt} = \frac{u}{\varepsilon}$ and $\frac{dy}{dt} = \frac{v}{\varepsilon}$, where ε is the porosity of the bed, equation (9) together with equations (1) and (2) yields

$$\frac{\partial h}{\partial t} = \frac{K}{\varepsilon} \left\{ \frac{\partial h}{\partial x} \frac{\partial \phi}{\partial x} - \frac{\partial \phi}{\partial y} \right\} \tag{12}$$

which describes the movement of the interface as a function of time.

The above equations were solved numerically using a finite-difference procedure. Details of this procedure are given elsewhere (Pinczewski and Tanzil (1980)).

Results from Computations and Experiments

    The drainage experiments were carried out on a two-dimensional model measuring
40 cm × 3.5 cm packed to a depth of 17 cm with glass beads having a diameter of 2 mm - see
Figure 3.   The bed was filled with distilled water and drained from a slit in the side
located 3 cm above the base of the bed.   The experiments consisted of draining the bed into
an evacuated vessel through a small orifice fitted in the flow line.   This arrangement
provided an almost constant liquid flow rate during the drainage process.   The position of
the interface as a function of time was recorded photographically.   For these experimental
conditions the hydraulic conductivity was 2.56 cm/sec and the effective porosity was 0.310
(obtained by draining the bed and measuring the collected liquid volume).

    Figure 4 shows a comparison between the computed and experimentally measured positions
of the gas-liquid interface during a typical drainage experiment.   The shape of the interface
is determined by the pressure gradients generated in the bed with the slope being greatest in
the vicinity of the drain point where the pressure gradient is greatest.   The overall
agreement between experiment and computation is good.   The differences between the computations
and experiment are within the range of uncertainty associated with the visual observations.

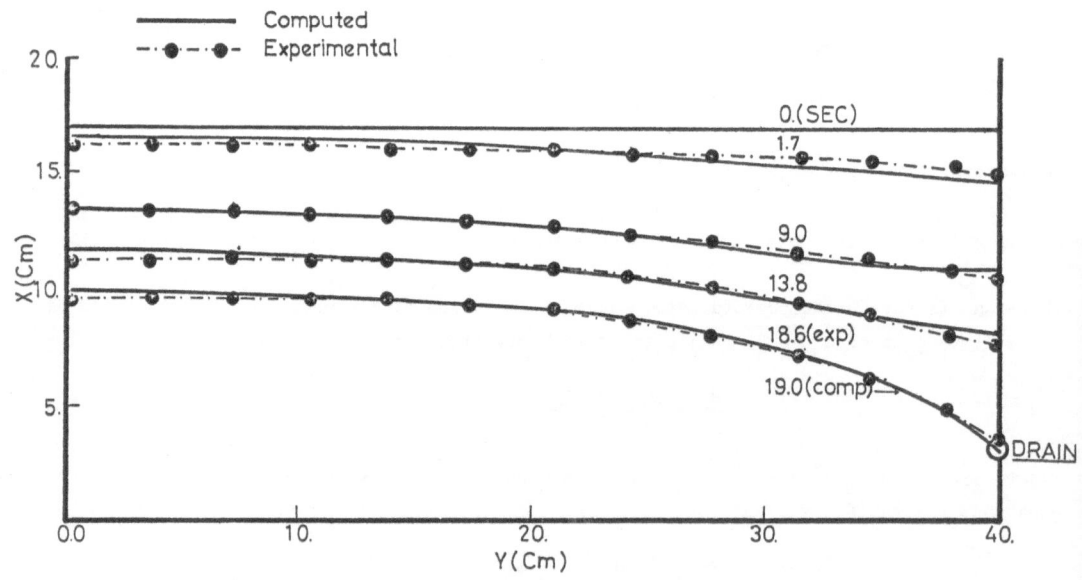

FIGURE 4 : Position of the gas-liquid interface during drainage (Ave. $Q = 20$ cm^3/sec,
        $K = 2.56$ cm/sec).

A more quantitative comparison of the results may be obtained by considering the time of arrival of the interface at the drain point. Table 1 shows this comparison together with a similar comparison for residual ratio which is defined as the ratio of liquid hold-up in the bed above the level of the drain when the interface arrives at the drain point to the initial liquid hold-up above the level of the drain point. The agreement between experiment and computation is excellent. Table 1 shows that increasing drainage rates result in increasing residual volumes of liquid.

TABLE 1 (K = 2.56 cm/sec)

Drainage rate	Time(s)		Residual Ratio	
(ml s^{-1})	Expt.	Computed	Expt.	Computed
11.1	41.2	42.5	0.347	0.341
19.6	19.3	19.7	0.458	0.452
35.3*	7.5	7.8	0.617	0.608

* average of three experimental runs, remainder average of two
  experimental runs.

The effect of hydraulic conductivity on residual ratio is shown in Figure 5. Again agreement between computation and experiment is good with the results showing that increasing hydraulic conductivity $(K = \frac{kg}{\nu})$ is associated with decreasing residual ratio, i.e. the higher the bed permeability and the lower the liquid kinematic viscosity the lower the residual ratio.

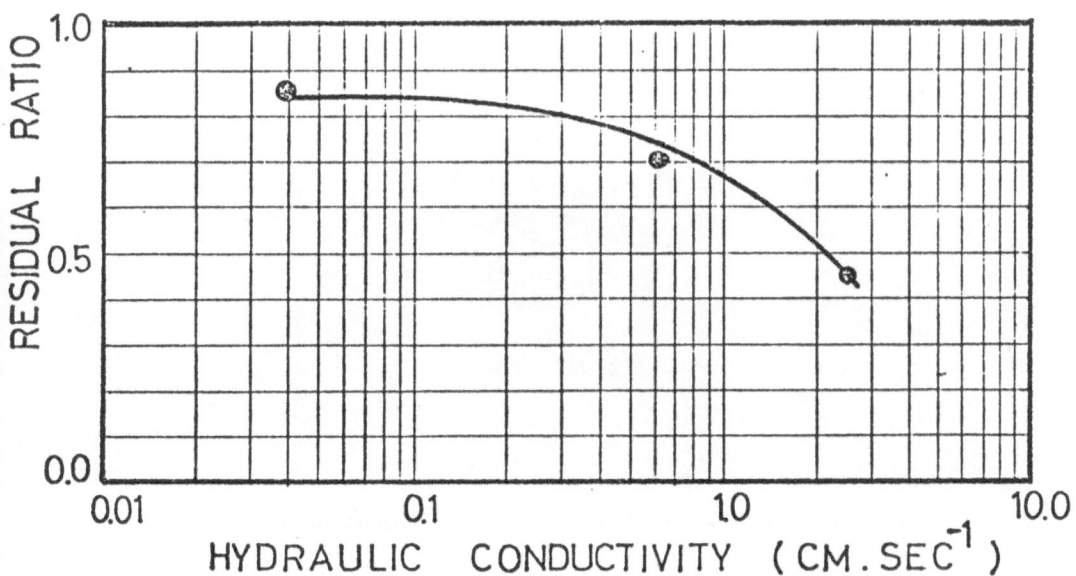

FIGURE 5 : Effect of hydraulic conductivity on Residual ratio : — computed;
● experimental (Q = 19.6 cm^3/sec).

The computed effect of permeability variations on bed drainage performance is shown in
Figures 6 and 7.   Both results are for a bed having a low permeability central region
surrounded by a higher permeability region.   Figure 6 shows the case of a tenfold variation
in permeability whilst Figure 7 shows the case of a one hundredfold variation.   For the case
shown in Figure 7, there is almost no drainage of liquid from the low permeability central
region.   In both of the cases shown the residual ratio is considerably higher than for the
case of a uniform bed having a permeability equal to that of the higher permeability regions
shown in Figures 6 and 7.

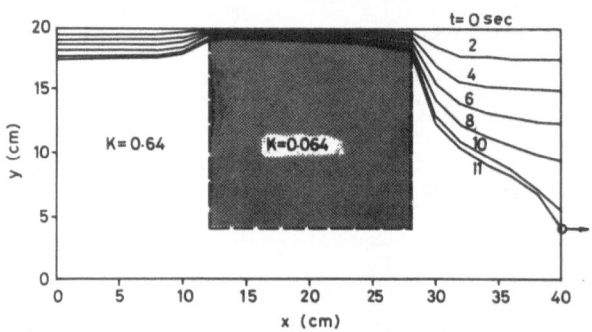

FIGURE 6 : Effect of variation in packing permeability (K ratio 10:1).

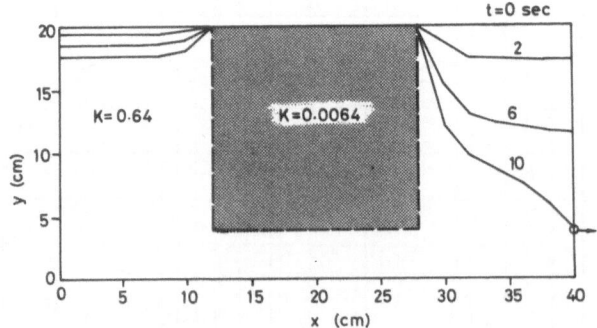

FIGURE 7 : Effect of variation in packing permeability (K ratio 100:1).

DISCUSSION

The simple two-dimensional model shows that the major factors which influence residual ratio are drainage rate and hydraulic conductivity. Hydraulic conductivity is in turn dependent on liquid viscosity and bed permeability. The residual ratio increases with increasing drainage rate and decreasing hydraulic conductivity. With regard to actual hearth drainage we can conclude that high tapping rates, high slag viscosities and low coke permeabilities will result in high slag residual volumes.

If the furnace is being tapped periodically, then the overall tapping rate can be reduced by tapping the furnace continuously and this will result in a decrease in residual slag volume. If the furnace is already being tapped continuously the pressure gradients in the vicinity of the taphole (or tapholes) may still be reduced by increasing the number of tapholes, i.e. by tapping simultaneously from two or more tapholes spaced around the circumference of the hearth, this will again reduce the residual slag volume and improve furnace performance.

Under typical Australian operating conditions, slag contains SiO_2 (35%), CaO (41%), Al_2O_3 (16%) and MgO (6%) on a weight percentage basis. At a fixed temperature slag viscosity is very sensitive to composition particularly to the amount of MgO present. Increasing the percentage MgO reduces slag viscosity and it is therefore possible to decrease residual volume by adding MgO to the furnace. The only disadvantage with this strategy is that the additive is rather expensive and the total slag volume flow is increased.

Residual slag volume may also be reduced by increasing the permeability of the coke bed. Poor quality low strength coke readily disintegrates as a result of the thermal, chemical and mechanical processes which occur in its passage through the furnace, and this results in a small mean coke particle size and high fines content in the hearth and a consequent reduction in bed permeability and porosity. This leads to high slag residual volumes and poor hearth drainage. Increasing the quality of the coke results in higher bed hydraulic conductivity and this leads to lower residual slag volumes and improved furnace performance.

All of the above strategies for reducing residual slag volumes and improving furnace performance have a cost penalty associated with them. Increasing coke quality, where this is possible, is extremely expensive. Modifying slag composition by means of additives and modifying tapping practice on an existing furnace are also expensive undertakings. The actual strategy(s) which should be adopted in any particular case will depend to a large extent on the range of options available and the results of a cost benefit analysis for the overall operation. A full three-dimensional simulation model for the hearth drainage process would be of considerable benefit in placing actual dollar values on the various options available to the furnace operator and this would constitute a major step in optimizing the overall furnace operation.

CONCLUSIONS

Poor hearth drainage is a serious problem in the operation of large blast furnaces. The problem is largely associated with the drainage of the slag phase and a clear need exists to improve hearth drainage.

On the basis of a simple two-dimensional mathematical model of the drainage process in packed beds, we conclude that hearth drainage may be improved by decreasing the average tapping rate i.e. continuously tapping the furnace from two or more tapholes, reducing slag viscosity and improving coke quality to increase coke permeability.   The most effective means of obtaining a realistic estimate of the relative cost benefit of any of these improvements to a large commercial blast furnace is by an extension of the simple two-dimensional drainage model to three-dimensions.   This would allow a full simulation of the drainage behaviour of an actual hearth in an operating blast furnace and therefore provide a true estimate of the effects of changes in operating conditions and raw material quality on overall furnace performance.

ACKNOWLEDGEMENTS

The assistance of Mr L. Nowak, B.H.P. Central Research Laboratories, in the measurement of the experimental data reported in this work is gratefully acknowledged.

REFERENCES

[1]   Burgess, J.M., Jenkins, D.R., McCarthy, M.J., Nowak, L., Pinczewski, W.V., and
         Tanzil, W.B.U. (1980)   'Lateral Drainage of Packed Beds'.   *Chemeca* 80, Proceedings
         of 8th Australian Chemical Engineering Conference, Melbourne, August 1980.

[2]   Burgess, J.M., Jenkins, D.R., and de Hoog, F.R., (1981)   'Mathematical Models for
         Gas Distribution in the Iron Making Blast Furnace'.   Proceedings of this
         Symposium.

[3]   Nakamura, N., Togino, Y., and Tateoka, M. (1978)   'Behaviour of Coke in Large Blast
         Furnaces'.   *Ironmaking and Steelmaking* 5 (1), 1.

[4]   Peaceman, D.W. (1977)   'Fundamentals of Numerical Reservoir Simulation'   Elsevier
         Scientific Publishing Company, 1977.

[5]   Peacey, J.G., and Davenport, W.G. (1979)   'The Iron Blast Furnace Theory and Practice'
         Pergamon Press, 1979.

[6]   Pinczewski, W.V., and Tanzil, W.B.U. (1981)   'A Numerical Solution for the Drainage
         of Two-Dimensional Packed Beds' *Chemical Engineering Science* (in press)

[7]   Strassburger, J.H. (1969) 'Blast Furnace - Theory and Practice' Gordon and Breach,
        New York, 1969.

[8]   Todsen, M. (1971) 'On the Solution of Solution of Transient Free Surface Flow
        Problems in Porous Media by Finite-Difference Method' Journal of Hydrology, <u>12</u>,
        177.

NOMENCLATURE

D	width of packed bed
g	acceleration due to gravity
h	height of gas-liquid interface
h_o	initial height of liquid in packed bed
k	bed permeability
K	hydraulic conductivity
P	pressure in liquid
Q	volumetric drainage rate of liquid from bed
q	volumetric outflow rate per unit volume of bed
t	time
t_o	time at start of drainage
u,v	local superficial velocity component in x,y directions
x,y	cartesian position co-ordinates
ε	porosity of packed bed
ρ	density of liquid
ϕ	hydraulic potential

A MATHEMATICAL MODEL FOR AN
URBAN DRAINAGE SYSTEM

R. VOLKER,
Department of Civil Engineering,
James Cook University of North Queensland, Townsville,

AND

P.C. TURL,
Design Engineering,
Townsville City Council, Townsville.

1. NATURE OF THE PROBLEM

Stormwater runoff from an area of approximately 2000 ha of the developed portion of Townsville drains to the tidal estuary of Ross Creek and thence to the sea. The connection of the catchment to Ross Creek is via a complex system of pipes, retention basins, lined and unlined open channels and culverts. At the outlet end the water flows through the Woolcock Canal which is a concrete channel of rectangular cross-section connected to Ross Creek by tide gates. As shown in Figure 1, the catchment is roughly bounded by Castle Hill at the north, Ross River at the east and south and by other natural drainage divides at the west and north west. A more detailed description of the physical system including an explanation of the difference between primary and secondary catchments will be given in section 2.

Following heavy rainfall in December 1976 the Woolcock Canal failed, causing extensive flooding, inundating property and houses in its lower reaches and disrupting traffic on a number of major roads. Another prolonged high intensity rain storm in January 1978 caused flooding similar to that in December 1976. As a result of these events the Townsville City Council instituted a reappraisal of the canal design with a view to identifying the major deficiencies and formulating proposals to rectify them. Although flooding occurred in some of the upstream sections of the drainage network the major problems occurred in the lower reaches and particular attention was directed to these areas.

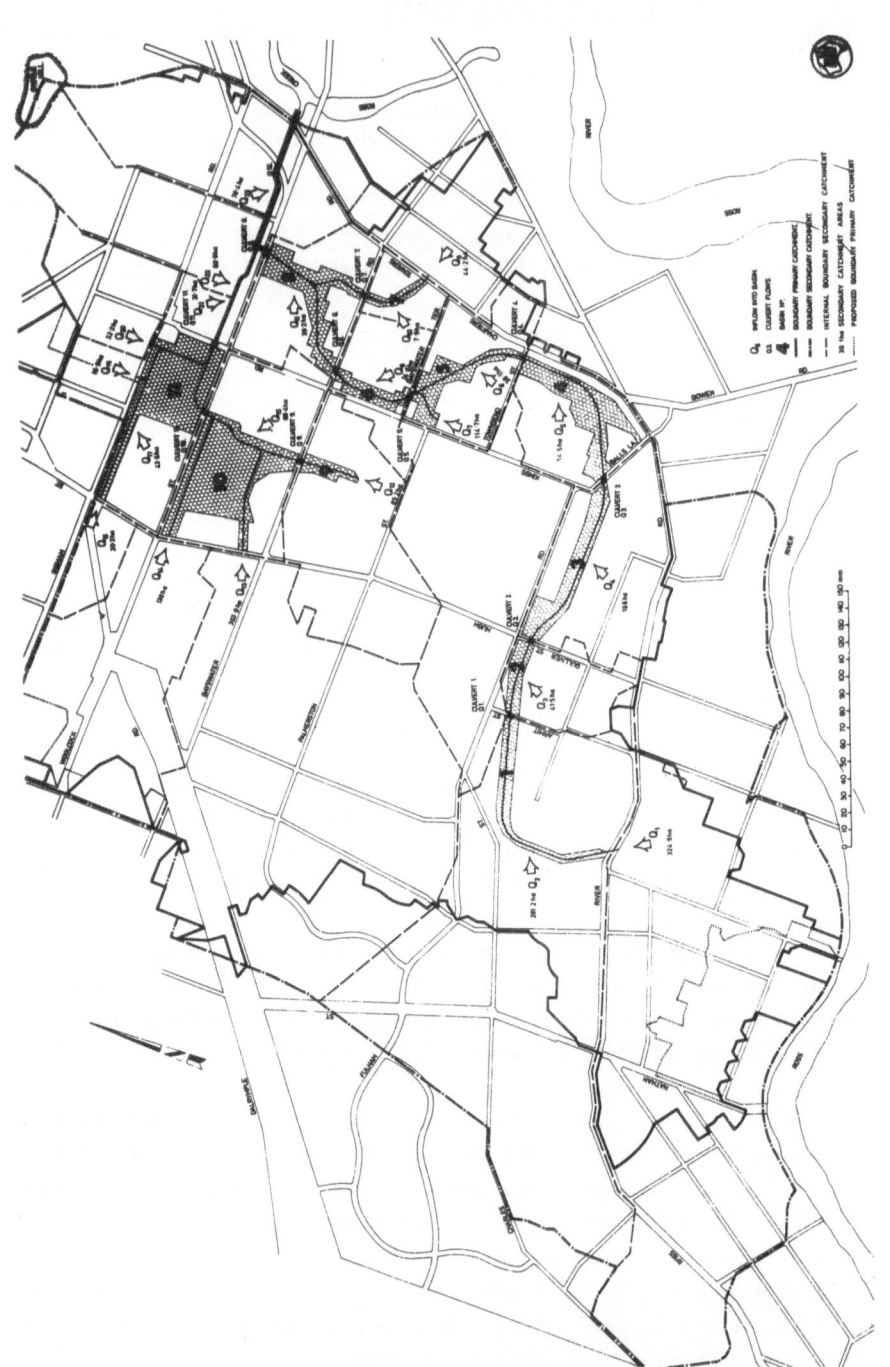

FIGURE 1 : Woolcock canal catchment and layout plan.

2. THE PHYSICAL SYSTEM

At the commencement of design the then existing drainage network consisted of a series of interconnected stormwater retention basins discharging eventually via concrete lined canals to Ross Creek. The basins were connected by culverts under major roadways.

2.1 Catchments

Figure 1 shows the catchment area contributing to the Woolcock Canal. The catchment can be broken up into two segments - the primary catchment and the secondary catchment. Primary drainage is defined as the system of underground pipes with inlets and other appurtenances, which drains a particular area. This area is therefore known as the primary catchment. Secondary drainage is the overland flow path followed by runoff waters when the primary drainage is surcharged. Open drains are usually paths of both primary and secondary drainage. The secondary catchment is the area which contributes inflows into a system when all primary drainage is surcharged.

Primary drainage often falls against the natural ground slope, to carry stormwater to the nearest suitable discharge point. This results in cases where the primary and secondary catchments do not coincide. For example, in Figure 1, the secondary catchment area contributing to inflow Q1 into retention basin No.1 is 324.9 ha, whereas the primary catchment area contributing to Q1 is considerably less than this because a substantial proportion of primary drainage from the catchment is piped against the natural fall of the land to discharge into Ross River. The boundaries of both primary and secondary catchments contributing to the Woolcock Canal are shown in Figure 1.

2.2 Retention basins

All of the inflows into the main drainage channel of the Woolcock Canal system, except inflows Q21 to Q23, enter into large depressed areas, before reaching the concrete lined sections of the canal. The depressions act as retention basins in that they store runoff for a period, resulting in the peak outflow from each basin being reduced, because the flow is spread over a longer period.

Basins 1 to 6 are the remains of a previous path of the Ross River, which has since moved south and east. Basin 7 was a slight depression which has been improved to form an open drain, reducing pondage and flooding in the adjacent area. Basins 8 to 11 are old areas of salt pans and mangroves, which, until the tide gates were constructed on the canal, were subjected to normal diurnal tidal flooding. All of the retention basins originally drained through a narrow tidal creek, located approximately on the same alignment as the main concrete lined canals.

A number of retention basins are used for purposes additional to drainage and retention storage. Basins 1 and 4 contain developed sporting fields, basin 3 lies within the new botanical gardens at Anderson Park, and basin 11 is an area in the city where trail bikes can be ridden.

2.3 Culverts

The downstream limit to each retention basin is defined by a constructed roadway, the outflow from the basin being controlled by the culvert under, or by overtopping of, the roadway The culvert barrels are reinforced concrete pipes of either circular or rectangular cross-section.   The discharge through the culverts and also the discharge over the roadway may depend on the water levels in both the upstream and downstream retention basins.

2.4 Concrete canals

The major investment in the Woolcock Canal system is approximately one million dollars (1972 prices) in the concrete lined canal in the lower reaches running from the outlet of culvert 8 to Ross Creek (see Figure 1).   This section has a depth of 2.44 m from the top of the concrete walls to the bed of the canal and a width of 9.25 m between walls.   The canal was designed to be surcharged by 0.6 m and for this purpose the area on either side was graded back from the top of the canal wall at a slope of 1 vertical to 5 horizontal giving the effective section shown in Figure 2.

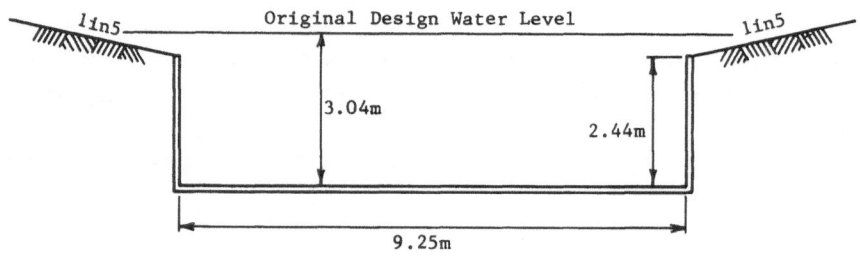

FIGURE 2 : Original design cross-section for the large canal.

The other section of concrete lined canal runs from the downstream end of retention basin 11 to join the main canal just downstream of the outlet from culvert 8.   This section had a design width of 3.18 m and a depth from the top of the wall of 2.44 m and was not designed for any surcharge above the walls.

It was obvious even at the preliminary design stage that a reduction in flood levels in the lower reaches would be achieved by an increase in the capacity of the main canals.   A check on the strengths of the concrete sections showed that the main canal could not be widened without major structural changes which would have made it necessary to rebuild the canal; however the smaller canal had been constructed in such a way and of sufficient strength that it could be doubled in width and still make use of the existing concrete wall and floor panels. This was a very significant factor to take into account when considering the options for improvement of the drainage network.

2.5 Tide gates

At the outlet of the main canal into Ross Creek is a radial type tide gate structure. The tide gate has a central supporting pier, and when closed rests on a 0.332 m hump in the canal cottom. The bottom hump, and the central pier cause energy losses in the flow.   Ross Creek is tidal with its outlet in Cleveland Bay, and although the creek carries stormwater from the City area and sections of other adjacent suburbs downstream the major proportion of its flow comes from the Woolcock Canal.

3. THE NEED FOR A MATHEMATICAL MODEL

Although the physical processes involved in the movement of rain water through the catchment and drains are reasonably well understood it is not yet possible within normal budgetary restraints to collect all of the data necessary to completely determine the fate of the water at any point in time.   It is also not normally economically feasible to solve all of the mathematical equations governing the individual processes.   Approximations are possible, however, which reduce the complexity of the problem to a manageable level and still give a reasonably accurate picture of the runoff and drainage processes.   The resulting equations will be discussed more fully in section 4.   The major processes involved will now be summarised.

Rainfall on the catchment may be intercepted on the vegetation, may infiltrate into the soil, may fill surface depressions or it may cause overland flow.   The particular catchment considered here is urbanised with the result that there is a relatively high percentage of its surface area which is impervious.   In this case experience shows that the rate of overland or piped flow into the main drainage channel from any of the subcatchments can be adequately described by an equation of conservation of mass of the water.   Thus from a knowledge of the subcatchment area, its slope and surface type and from a knowledge of the rainfall intensity and duration, an equation can be obtained which describes the inflows to each retention basin (Q1, Q2, etc. in Figure 1) as a function of time.

In each retention basin the water contributes to an increase in storage in the basin. There is also an exchange of water between the basins depending on the relative levels on each side of the connecting culverts.   The water level in each basin and the outflow from it can be obtained by solving the equation of mass conservation of the water entering, leaving, and being stored in, the basin.   This solution, however, depends on the equation governing flow in the culverts which in turn depends on the water levels in upstream and downstream basins. The culvert flow equations are derived from considerations of energy losses in the pipes and at their entrances and outlets.   Because of the interdependence of the water levels in the basins the equations are usually solved numerically by discretizing with small time intervals.

Discharge through the concrete lined canal depends on the properties of the canal such as slope, cross-sectional area and shape, and surface roughness as well as on the water levels and velocities at the upstream and downstream ends.   A complete description of the unsteady

flow in the canal requires a solution of the mass and momentum conservation equations.
Analytical solutions for the most general case are not available and one must resort to
numerical solutions.   When the changes in water volume stored in the canal are small
compared with the volumes passing through the canal the governing equations can be simplified
and this is the case for the Woolcock canal.   Nevertheless the water levels and velocities
at the upstream and downstream ends of the canal are time dependent.   The downstream
conditions are governed by tidal levels in Ross Creek while the upstream conditions are
governed by the water surface elevations in the retention basins relative to those in the
canal.   Iterative numerical solutions of the governing equations are therefore employed for
canal flows.

It can be seen that although the individual mathematical equations to be solved are
relatively simple, the interdependence of the boundary conditions renders the solution of the
entire system significantly more difficult.   To design an improved drainage system one needs
to be able to calculate, for any given rainfall event, the resulting water levels in and
adjacent to the main channels.   The mathematical equations describing the individual
processes and taking account of the links between them enable these calculations to be
performed.   For the system described the only feasible way of obtaining the results is to
use a digital computer to solve the governing equations and produce flood levels as a function
of time and position in the drainage system.

An alternative method of design may have been to use a scaled physical model of the
drainage network.   Such a model of the system discussed here would have been costly to
construct and operate.   Physical models are not free from approximations because of the
different scaling requirements to account for different processes and because of the limited
accuracy of reproduction of actual topography.   There is also less flexibility to test
alternative remedial measures.   Consequently the decision was made to develop a mathematical
computer model to predict flood levels to assist in the redesign of the drainage system.

4.   THE GOVERNING EQUATIONS

4.1   Catchment runoff

The total catchment area is divided into 32 subcatchments for each of which the area
was determined and the percentage of rainfall that produces direct runoff was estimated.   The
maximum time taken for the runoff to reach the subcatchment discharge point was also estimated;
this is usually called the time of concentration.   It was assumed that the area contributing
to runoff varied linearly with time up to the time of concentration.   Other assumptions were
tested but because of the shape and slope of the catchments the linear case proved satisfactory
Consider the direct runoff from a portion A of the subcatchment shown in Figure 3 where t_1
and t_2 are flow times to the retention basin (or canal for subcatchments in the lower
reaches).

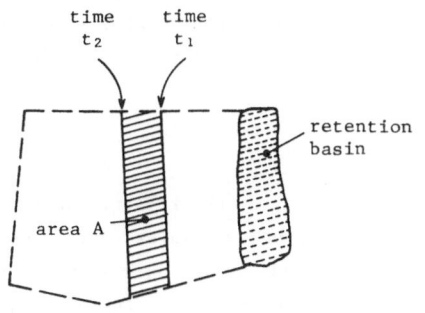

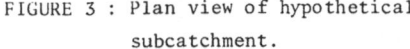

FIGURE 3 : Plan view of hypothetical
          subcatchment.

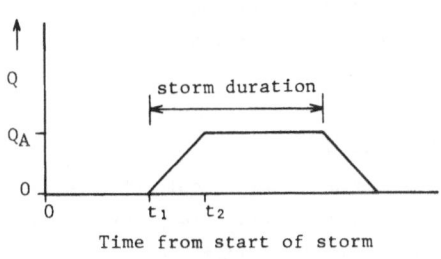

FIGURE 4 : Discharge from area   A.

If the intensity of rainfall on portion A is I mm hr^{-1} then the direct runoff from
that area is calculated from :

$$Q_A = \frac{CIA}{3.6 \times 10^6}$$

(1)

where Q_A is discharge $[m^3 s^{-1}]$;   A is the area $[m^2]$;   and C is a dimensionless runoff
coefficient which represents the fraction of rainfall which produces direct runoff.   The way
in which the discharge from the portion A varies with time for a constant intensity rainfall
is shown in Figure 4.   To determine runoff from the whole subcatchment the contributions from
all areas such as A are summed taking account of the appropriate times of flow.   To allow
for variations in intensity of rainfall the total storm duration is divided into a number of
smaller intervals during each of which the intensity is approximately constant.   Thus by
dividing the catchment into a number of equal areas and by dividing the storm duration into
smaller time intervals, determining the runoffs and superimposing them with due allowance for
time differences, a reasonable representation of the runoff hydrograph can be obtained for a
variable intensity storm on the whole catchment.   The size of the subareas and the lengths of
the time intervals are chosen to provide suitable accuracy.

4.2 Routing through retention basins

When water is flowing into a reservoir at one point and out at another the change in
volume stored will depend on the relative rates of inflow and outflow.   For the retention
basins inflow will generally come from the surrounding subcatchment and from the upstream
culvert and outflow will occur through the downstream culvert.   The governing equation is :

$$Q_I - Q_O = \frac{dS}{dt}$$

(2)

where Q_I is inflow to the basin $[m^3 s^{-1}]$; Q_O is outflow from the basin $[m^3 s^{-1}]$;
S = volume stored in the basin $[m^3]$;   and t is time [s] .

In equation (2) both Q_I and Q_0 depend to some extent on discharge through the culverts and these in turn depend on water levels in the retention basins (upstream, downstream and the one under consideration).   The storage S is a function of water level and the basin geometry; this function was evaluated from survey information and a curve fitted to the result for each basin.

4.3 Culvert flow

Discharge through a culvert depends, amongst other factors, on the relative water levels upstream and downstream.   In general a distinction can be drawn between inlet control when the tailwater (downstream) level does not affect culvert discharge and outlet control when the tailwater level and the properties of the culvert pipe do affect discharge.   Another criterion of importance is whether the entrance is submerged or not.   For the situation shown in Figure 5 where H is water depth [m] above the bottom of the pipe at the entrance, experiments indicate that the entrance will be unsubmerged provided H/D < 1.2 .

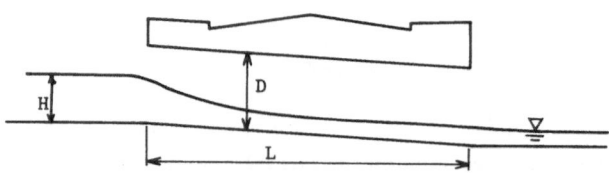

FIGURE 5 : Culvert with unsubmerged entrance.

For a circular pipe of diameter D[m] the discharge $Q[m^3 s^{-1}]$ is given by the following approximate relations (Henderson, [1])

$$0 < H/D < 0.8 \qquad \frac{Q^2}{D^2\sqrt{gD}} = 0.48 \left(\frac{H}{D}\right)^{1.9} \tag{3}$$

$$0.8 \le H/D < 1.2 \qquad \frac{Q^2}{D^2\sqrt{gD}} = 0.44 \left(\frac{H}{D}\right)^{1.5} \tag{4}$$

where g is gravitational acceleration $[m^2 s^{-1}]$.

A somewhat simpler expression applies for a pipe of rectangular section.

For inlet control and a submerged entrance (H/D > 1.2) the discharge through a circular pipe of diameter D is :

$$Q = C_d \frac{\pi}{4} D^2 \sqrt{2g\left(H - \frac{D}{2}\right)} \tag{5}$$

where C_d = discharge coefficient (dimensionless).

For outlet control as depicted in Figure 6, the velocity V in the culvert is obtained from :

$$\left(H_1 + z_1 + \frac{V_1^2}{2g}\right) - \left(H_2 + z_2 + \frac{V_2^2}{2g}\right) = \left(K_e + \frac{fL}{4R} + K_0\right)\frac{V^2}{2g} \tag{6}$$

where subscripts 1 and 2 refer to upstream and downstream conditions respectively; z is the elevation of the channel bed above a horizontal datum [m]; V is water velocity [ms^{-1}]; K_e , K_0 are entrance and outlet energy loss coefficients [dimensionless]; f is pipe friction coefficient [dimensionless]; L is length of pipe [m]; R is the hydraulic radius [m] which is area of cross-section divided by wetted perimeter of the pipe.

If outlet control applies and the pipe does not flow full at the downstream end the solution of equation (6) is not straightforward but this case is unimportant in the present study because, for the critical design flows, the pipes flow full.

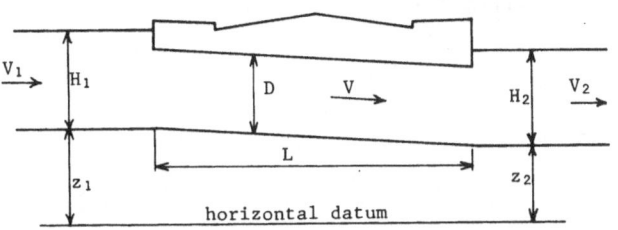

FIGURE 6 : Culvert flowing full.

For flood conditions water often overtops the roadway above the culverts.   This flowrate is calculated by analogy with a broadcrested weir taking due account again of whether the tailwater controls the flow or not.

4.4 Canal flow

The water velocity through the canal is calculated from the empirical Manning equation (Henderson, [1]):

$$V = \frac{R^{2/3}S^{\frac{1}{2}}}{n} \tag{7}$$

where R is the hydraulic radius defined previously; S is the slope of the total energy line [dimensionless]; and n is an empirical coefficient.

The slope of the energy line depends on the water levels and velocities at the tide gates downstream and on the culvert inflows from upstream.

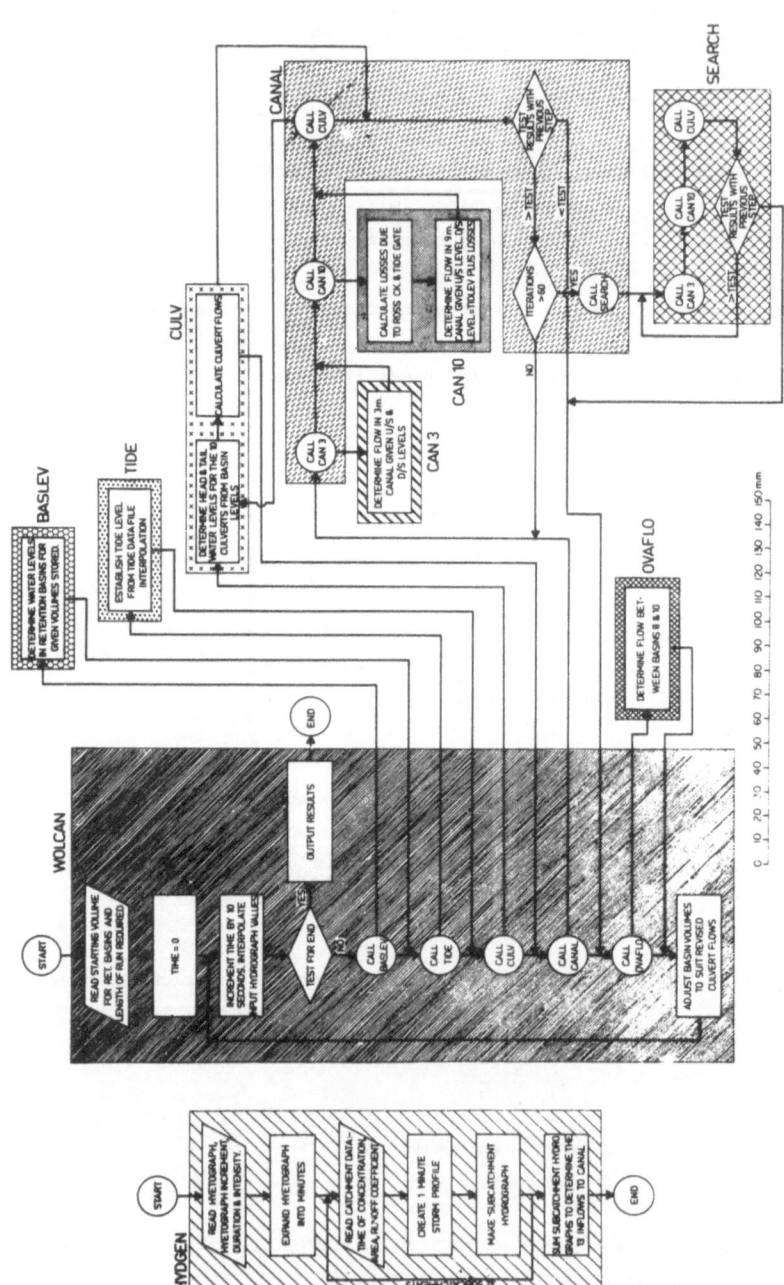

FIGURE 7 : Woolcock canal computer model flow chart.

5. MODEL FORMULATION

The computer model for the drainage system consists of two main programs and eight subroutines as shown in Figure 7.   A detailed description of each section has been given in a report by Townsville City Council [2] and only a summary is possible here.   One of the main programs, HYDGEN in Figure 7, generates runoff hydrographs for each of the subcatchments while the other WOLCAN, routes the runoff through the retention basins and canals.

5.1 Hydrograph generation

Rainfall information in the form of a hyetograph (a plot of rainfall intensity versus time) is required as well as relevant catchment data.   The total rainfall duration is divided into a number of consecutive shorter duration events, each usually 1 hour for the design storms. Each subcatchment was divided into subareas corresponding to one minute differences in flow time to the nearest retention basin or canal.   The total runoff hydrographs are generated as explained in section 4.1 by the program HYDGEN according to the procedure shown in Figure 7.

5.2 Routing through the drainage channels

The basic flow chart for the main routing program, WOLCAN, is shown in Figure 8.   It calls several subroutines and the detailed arrangement is given in Figure 7.   All necessary data are read in and the initial conditions are prescribed;   the data include the runoff hydrographs generated by HYDGEN for the particular storm being modelled.   The routing is carried out using an explicit procedure for integration with time in the routing process.   A small time interval, usually 10 seconds, is used and the loop in Figure 8 is repeated for every interval.

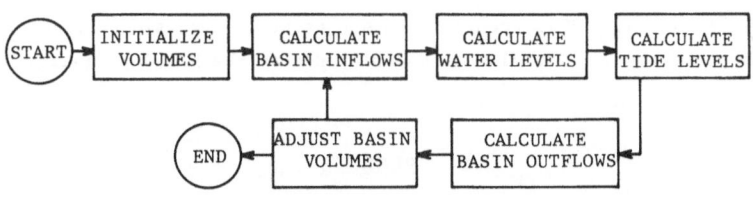

FIGURE 8 : Basic flow chart for WOLCAN.

The subroutine BASLEV determines the basin water level for a given storage volume including upstream and downstream water levels for calculating culvert flows.   Water surface slopes in the basins are determined but these are small except at the start of the storm.

Subroutine TIDE calculates the tide level at the outlet using a sinusoidal relation between height and time for prescribed low and high water levels.   The subroutine CULV determines culvert flows out of retention basins 1 to 10 according to the principles outlined in section 4.3.   The subroutine can cater for flows in a reverse direction through the culverts when necessary and also for flow over the roadways when a culvert overtops.

The flows in culvert 8 and in the two concrete lined canals are determined by sub-routine CANAL. For any particular time-step there are three known water levels (in basins 8 and 11 and in Ross Creek) and known catchment inflows into the smaller (3 m) canal and the upstream end of the larger (9 m) canal. The problem solved by CANAL is schematically illustrated in Figure 9.

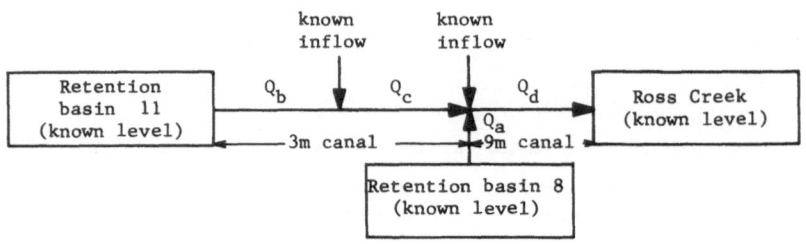

The unknowns are the water level at the canal junction and the flows in culvert 8, in the upper and lower halves of the smaller canal and in the larger canal. These flows are represented by Q_a , Q_b , Q_c and Q_d respectively in Figure 9. Subroutine CANAL uses an iterative process to solve for the unknowns. It calls subroutines CAN3 and CAN10 to determine flows in the small and large canals respectively and also calls SEARCH to assist the convergence procedure when necessary. The convergence test is for continuity of flow at the junction of the canals. An interval halving technique is normally used to obtain the correct flows but for certain conditions the flows are sensitive to small changes in levels and the subroutine SEARCH is then employed to achieve convergence.

Another subroutine used in the routing is OVAFLO which models overland flow between retention basins 8 and 10. This occurs through a natural depression between the basins and can be in either direction. The flow is calculated from the Manning equation (see section 4.4).

6. RESULTS

To check the assumptions made in formulating the model and to calibrate the parameter values, it was used to simulate runoff resulting from a rainstorm associated with Cyclone 'Keith' on 30th and 31st January 1978. Maximum flood levels for this storm had been recorded by Townsville City Council surveyors and two continuous rainfall records were available for sites close to the catchment. The storm duration was 23 hours and the average rainfall intensity was 12.24 mm hr^{-1} for catchments contributing to basis 1,2,3,9,10 and 11 and 13.17 mm hr^{-1} for the remainder. Table I compares field records of maximum water levels in various basins with values obtained from the model after calibration.

Table I   Observed and Modelled Flood Levels for Cyclone 'Keith'

Basin	1	2	3	4	6	8	10	11
Field level (m)	6.73	6.48 5.59	5.21	4.28 3.96	>3.50	3.15	3.48 3.51	3.32 3.49
Model level (m)	6.23	5.58	4.95	4.13	3.73	3.32	3.50	3.48

Some of the field values were estimates because they were obtained from debris levels and alternative estimates from different sources have been shown in Table I where the true record is not clear.

To check the calibrated parameters the model was applied to a second storm of 5th and 6th January 1980 and the agreement between recorded and measured levels was very satisfactory.

The model was then used to investigate alternative remedial measures to reduce flood levels.   Storms of different durations and intensities were tested to determine the critical one for a design recurrence interval of 50 years in conjunction with a maximum tide level of mean high water springs.   The design storm adopted was of 6 hours duration with an average rainfall intensity of 29.25 mm hr^{-1} .   This storm on the existing system produced significantly higher flood levels than the longer duration events of January 1978 and January 1980. No flood level records were available for a comparable six hour or similar short duration storm of high intensity.

Several improvements to the drainage system were suggested.   These were :

(1)   widening the small canal to 6 m in width;

(2)   enlarging culvert 11;

(3)   streamlining the junction of the canals with culvert 8;

(4)   increasing the storage capacities of the retention basins by 571,000 m^3;

(5)   streamlining the tide gate structure;

(6)   piping of extra primary drainage out of the system;

(7)   elimination of the flow path between basins 8 and 10.

The model was run for the design storm with all of these modifications and a comparison of flood levels with those for the existing system is given in Table II.

Table II   Comparison of Peak Flood Levels from the Model

Retention basin	1	2	3	4	5	6	7	8	9	10	11
Peak level- existing system (m)	7.19	7.11	5.14	4.45	4.36	4.25	4.26	4.12	4.23	4.24	4.25
Peak level- all modifications (m)	6.16	5.72	5.07	4.09	3.79	3.76	3.29	3.19	3.21	3.19	3.12

The reductions in flood levels are very significant.   To help determine which modifications produce the most marked improvement the model was run with each change separately and
with several combinations of different changes.   This enables a cost-benefit comparison of the
alternatives.   It also demonstrates the flexibility of the mathematical model and its value
as a design tool.

7. CONCLUSIONS

An urban catchment draining through a series of retention basins to concrete lined
canals which discharge to a tide affected estuary presents major problems for designers seeking
to improve the system and reduce flood levels.   Mathematical equations describing the various
processes can be used to determine the flood levels for any chosen storm events.   However,
because of the large number of components in the drainage network and because of the interdependence of flows and water levels in adjacent components, the only feasible way to solve
the equations is with a mathematical model programmed for a digital computer.   This model
simulates the runoff and routing of water through the catchment at discrete time intervals and
produces flood levels as a function of time at points throughout the system.   It has been
shown that such a model is an invaluable aid in choosing appropriate measures to reduce flood
levels to acceptable values.

8. ACKNOWLEDGEMENTS

The authors express appreciation to their respective organisations which provided the
opportunity to undertake the work described and to present the results.

9. REFERENCES

[1]   Henderson, F.M. *Open Channel Flow*, Ch.4.   The Macmillan Company, New York, 1966.

[2]   Townsville City Council, *Woolcock Canal Design Report*, 1980.

APPLICATION OF ACTUARIAL TECHNIQUES TO NON-LIFE INSURANCE:
ESTABLISHMENT OF PROVISIONS FOR OUTSTANDING CLAIMS

G.C. TAYLOR,

E.S. Knight & Co., Consulting Actuaries, Sydney.

1. INTRODUCTION

A major problem facing management, particularly in the "long-tailed" liability classes, is the determination of a suitable *provision (or reserve) for outstanding claims*. The need for the provision arises from the delays (often substantial) which occur :

- (i) between the occurrence of the event generating the claim and notification of the claim to the insurer;
- (ii) between notification of the claim and its final settlement.

It is almost universally accepted that the insurer incurs a liability at the time of occurrence of the event generating the claim. Between this date and the date of final settlement, i.e. during delays (i) and (ii) above, the insurer has an unpaid liability on his hands. He must establish the necessary provision in his accounts.

In this paper a variety of methods of estimating outstanding claims is considered. Most of the methods are currently in practical use. In the case of each method some details of goodness-of-fit of the data to the underlying model are given. With a couple of exceptions, the goodness-of-fit does not appear to differ greatly from one method to another. Yet estimates derived from the same data set range from $12.8M to $25.0M - variation by a factor of nearly 2.

It is clearly of the utmost financial importance for an actuary to be able to resolve between the alternatives.   The matter could conceivably be one of life and death for the insurer.

In later sections of this paper, some features of the wide variation are considered. A means of overcoming the difficulties inherent in the standard methods is presented.

It is apparent, however, from what has been said already that the problem faced here is one in which skill in mathematical modelling is critical.   Models which are ostensibly not too different produce wildly different results.   The mathematical techniques used in the modelling are all elementary.   The crucial factor appears to be design of the model rather than sophistication in fitting it to the data.

2.   FORM OF DATA

The form of data considered in this paper is standard.   It is more or less the form which would be built up by the statistical returns prescribed by statute in both Australia and the U.K.

Claims are recorded by *year of origin*, i.e. the office year in which the event generating the claim occurred.   It is common in Australian terminology to refer to this as the *accident year*.   Claim payments in respect of each accident year are subdivided according to *development year* of payment, where development year j is the office year given by :

$$\text{accident year} + j .$$

Thus the primary set of data on which analysis is based takes the following form :

$$
\begin{array}{lllll}
c_{00} & c_{01} & c_{02} & \cdot\ \cdot\ \cdot \\
c_{10} & c_{11} & c_{12} & \cdot\ \cdot\ \cdot\ ; \\
\cdot & \cdot & \cdot \\
\cdot & \cdot & \cdot \\
\cdot & \cdot & \cdot
\end{array}
\qquad (2.1)
$$

in short, an array of recorded values,

c_{ij} = amount of claim payments in respect of year of origin i and development
year j .

Note that, for convenience, the earliest year of origin entering into the data has been labelled 0 .

It is usual in practice for the array (2.1) to assume some regular shape, but this is not a prerequisite of the methods of analysis considered in this paper.

A concomitant of the array (2.1) is another array :

$$
\begin{array}{cccc}
n_{00} & n_{01} & n_{02} & \cdot \quad \cdot \quad \cdot \\
n_{10} & n_{11} & n_{12} & \cdot \quad \cdot \quad \cdot \; ; \\
\cdot & \cdot & \cdot \\
\cdot & \cdot & \cdot \\
\cdot & \cdot & \cdot
\end{array}
\qquad (2.2)
$$

where

n_{ij} = the number of claims, in respect of year of origin i , finalized in development year j .

Define

$$
n_i = \sum_{j=0}^{\infty} n_{ij}
$$

= total number of claims incurred in year of origin i .

Some variants of the methods described in this paper define n_{ij} differently. For example, Sawkins [7], [8] discusses the use of *number of claims handled* rather than number of claims finalized. Taylor [9] considers wider options. We defer discussion of the points arising out of this choice until Section 14. Suffice to say at this juncture that throughout this paper n_{ij} will be used as defined above.

3. THE GENERAL STRUCTURE

In many forms of insurance, and indeed in the liability classes considered here, the amount of a claim will be directly affected by inflation. And in fact the claim amount will usually be related to dollar values *at the time of payment of the claim* rather than at the time when the claim was incurred. Therefore, a general model of the claim payment process is as follows.

Let C_{ij} , N_{ij} denote the random variables whose realizations were denoted in Section 2 by c_{ij} , n_{ij} respectively. Then

$$
E[C_{ij}] = n_i \mu_i \rho_{ij} \lambda_{ij} , \qquad (3.1)
$$

where

λ_{ij} is the value of an *inflation index* applicable to development year j of year of origin i ;

μ_i is the *average claim size* in year of origin i conditional on $\lambda_{ij} = 1$ for all j , i.e. it is the *deflated average claim size* ;

ρ_{ij} is the expected proportion of μ_i payable in development year j , still on the assumption that $\lambda_{ij} = 1$ for all j , i.e. it describes the *distribution of payment delays* ;

where stochastic independence of the four variables represented on the right side of (3.1) has been assumed.

This structure is discussed by Johnson [3]. Similar structures were considered by Finger [2] and Matthews [4].

In the majority of applications, it is assumed that

$$\rho_{ij} = \rho_j ; \tag{3.2}$$

$$\lambda_{ij} = \lambda_{i+j} . \tag{3.3}$$

That is, the delay distribution is independent of year of origin; and the inflation index depends only on the year in which payment is made, and inflation affects all development year equally.

Hitherto, only one application (Reid [5]) has adopted a more general assumption than (3.3). In most practical situations, insufficient data are available for the estimation of the number of parameters which arise from the generalization of (3.3). However, some generalizations of (3.2) will be considered in this paper.

Note that the information n_{ij} does not appear in (3.1). Clearly, it is possible to model ρ_{ij} as dependent on $\{n_{ij}\}$.

4. DATA

The data used in this exercise are presented in Appendix A. They consist of the n_i , the c_{ij} , the n_{ij} , the λ_{ij} , and values of c_{ij}/λ_{i+j} . Note, from (3.1), that

$$E[C_{ij}/\lambda_{ij}] = E[C_{ij}/\lambda_{i+j}] \qquad \text{where (3.3) holds}$$

$$= n_i \mu_i \rho_{ij} , \tag{4.1}$$

if λ_{ij} is regarded as nonstochastic.

It may also be noted at this point that, if n_i is regarded as known, then

$$E[C_{ij}/n_i\lambda_{ij}] = \mu_i\rho_{ij} \ . \tag{4.2}$$

The data are drawn from a Compulsory Third Party (CTP) portfolio.

5. METHODOLOGY

The following sections examine each of seven methods of analysis. In respect of each method the following steps are taken :

(i) the method is described;
(ii) the relation between the method under consideration and the general equation (3.1) is elucidated;
(iii) the parameters involved are estimated;
(iv) the fit of the model produced by these estimates to the data is examined;
(v) outstanding claims are projected.

In Step (iv) summary statistics of goodness-of-fit are *not* produced. Emphasis is rather placed on a critical scrutiny of goodness-of-fit in more detail. Indeed, it will be argued in Section 14 that such an approach is essential; that the use of summary statistics is courting danger.

The main statistics relied on in examination of goodness-of-fit are comparisons of actual and fitted;

(i) row totals (year of origin totals);
(ii) column totals (development year totals);
(iii) diagonal totals (payment year totals);

of claim payments in tables of the same format as Appendices A3 and A5.

6. METHOD 1 - BASIC CHAIN LADDER

6.1 Method

This is one of the first actuarial methods. Its evolution is not all that clear from the literature. To the author's knowledge, the method grew up in the U.K. Department of Trade under the auspices of R.E. Beard. Its use by this Department is documented by Sawkins [6, p.353].

On a superficial view, the method ignores inflation. However, effectively it is shown by Taylor and Matthews [10, pp.197-8] that the method can be regarded as assuming the rate of inflation constant over time :

$$\lambda_{ij} = \alpha^{i+j} , \tag{6.1.1}$$

for some constant α .

The method further adopts the simplification (3.2), whence the general structure (3.1) reduces to :

$$E[C_{ij}] = n_i \mu_i \rho_j \alpha^{i+j} \tag{6.1.2}$$

Define

$$A_{ij} = \sum_{k=0}^{j} C_{ik} \tag{6.1.3}$$

$$= \text{accumulated claim payments to the end of}$$
$$\text{development year } j , \text{ in respect of year}$$
$$\text{of origin } i .$$

From (6.1.2) and (6.1.3),

$$E[A_{ij}] = n_i \mu_i \alpha^i \sum_{k=0}^{j} \rho_k \alpha^k . \tag{6.1.4}$$

The method consists of forming ratios :

$$M_{j,j+1} = \sum_{i \in S} A_{i,j+1} / \sum_{i \in S} A_{ij} , \tag{6.1.5}$$

where S is the maximal set of values of i such that both numerator and denominator are functions of the array of available information.

From (6.1.4) and (6.1.5), it follows that, for large samples,

$$E[M_{j,j+1}] = \sum_{k=0}^{j+1} \rho_k \alpha^k / \sum_{k=0}^{j} \rho_k \alpha^k . \tag{6.1.6}$$

The estimates $M_{j,j+1}$ are "chained" :

$$M_{jk} = M_{j,j+1} M_{j+1,j+2} \cdots M_{k-1,k} , \quad \text{provided } k > j . \tag{6.1.7}$$

For large samples,

$$E[M_{jk}] = \sum_{h=0}^{k} \rho_h \alpha^h / \sum_{h=0}^{j} \rho_h \alpha^h . \tag{6.1.8}$$

In practice the above procedures will produce values of the M_{jk} up to a maximum value of K of k , this maximum depending on the extent of the array of data (2.1). Conceptuall what are required are estimates of the factors :

$$\sum_{k=0}^{\infty} \rho_k \alpha^k \Big/ \sum_{k=0}^{j} \rho_k \alpha^k = \left(\sum_{k=0}^{K} \rho_k \alpha^k \Big/ \sum_{k=0}^{j} \rho_k \alpha^k \right) \times \left(\sum_{k=0}^{\infty} \rho_k \alpha^k \Big/ \sum_{k=0}^{K} \rho_k \alpha^k \right) \qquad (6.1.9)$$

By (6.1.8), an asymptotically unbiased estimator of (6.1.9) is :

$$M_{j\infty} = M_{jK} M_{K\infty} \qquad (6.1.10)$$

where $M_{K\infty}$ is an asymptotically unbiased estimator of the second factor of (6.1.9) and must be obtained from some source exogenous to the data.

Outstanding claims are now projected as follows.   For large samples,

$$E[A_{i\infty}] = E[A_{if_i}] E[M_{f_i\infty}] , \qquad (6.1.11)$$

where development year f_i is the latest one observed in respect of year of origin i .   It then follows that :

$$E[A_{i\infty}] - E[A_{if_i}] = E[A_{if_i}]\{E[M_{f_i\infty}]-1\} . \qquad (6.1.12)$$

The interpretation of this equation is :

estimated outstanding		total claim payments		factor in
claims for year of	=	observed	×	braces.
origin i				

6.2 Estimates of parameters

If m_{ik} denotes the observed value of the random variable M_{jk} , then the following results are derivable from Appendix A3.

j	$M_{j,j+1}$ (a)	$M_{j\infty}$ (b)
0	4.04791	17.47849
1	1.99158	4.31790
2	1.37287	2.16808
3	1.21488	1.57923
4	1.10073	1.29991
5	1.06658	1.18095
6	1.05070	1.10723
7	1.01513	1.05380
8		1.03810

Notes : (a) *From equation (6.1.5).*

       (b) *From equation (6.1.10). The value of $M_{K\infty} = M_{8\infty}$ has been taken as 1.03810. This value, when incorporated in (6.1.12), produces a value of \$0.1M as estimated outstanding claims in respect of accident year 1971. This is approximately the amount paid in development years 9 and 10 of this accident year.*

       In this choice of $M_{8\infty}$, it has been taken implicitly that $M_{10,\infty} = 1$ This is not realistic, but does assist in a comparison of the estimated outstanding claims by this and later methods.

6.3 Goodness-of-fit

A table of actual claim payments, and the corresponding fitted payments is given in Appendix B1.

The process by which fitted claim payments have been obtained causes identity of total actual and fitted claim payments in each accident year (see Note (b) to Appendix B1). Moreover, the method of estimation of parameters (Section 6.1) causes identity of total actual and fitted claim payments in each development year.

A comparison of actual and fitted payments by payment year is as follows :

Payment year	Claim payments (a)		
	actual (b)	fitted (b)	deviation (c)
	\$	\$	%
1969	57369	103202	- 44
1970	636257	464382	+ 37
1971	1190620	1026788	+ 16
1972	1768638	1573515	+ 12
1973	1852004	2113115	- 12
1974	1879159	2602605	- 28
1975	3180205	2996728	+ 6
1976	3167312	3324248	- 5
1977	4206572	3642775	+ 15
1978	4038353	4009115	+ 1
1979	4415120	4535135	- 3

Notes : (a) *For payment year shown, but only in respect of the accident years included (for that payment year) in the basic data (Appendix A3).*

       (b) *From Appendix B1.*

       (c) *(actual - fitted)/fitted.*

6.4 Projected outstanding claims

Outstanding claims are projected according to (6.1.12), with the following results.

Accident year	Estimated outstanding claims at 31/12/79 (a)
	$M
1971	0.100
1972	0.167
1973	0.376
1974	0.564
1975	0.910
1976	1.521
1977	2.822
1978	4.253
1979	6.992
Total (1971-1979)	17.705(b)

Notes : (a) *Recall that, strictly, these estimates disregard claim payments in development years 10 and higher (see Note (b) to Section 6.2).*

(b) *This figure is not directly comparable with later projected totals of outstanding claims.   See Section 6.5 for comment.*

6.5 Comment

The main feature of the basic chain ladder method which distinguishes it from other methods is the assumption (6.1.1) of a constant inflation rate.   The actual inflation rates observed over the period of claims experience are as follows :

Year	AWE inflation rate (a)
	%
1969 to 1970	9.2
1970 to 1971	15.4
1971 to 1972	4.8
1972 to 1973	12.0
1973 to 1974	22.4
1974 to 1975	19.7
1975 to 1976	15.3
1976 to 1977	9.6
1977 to 1978	8.6
1978 to 1979	7.8
Average (b)	12.3

Notes :   (a)   *Derived from Appendix A4.*

(b)   *This is not an arithmetic average.   It is derived from Appendix A4 as :*

$$(0.952377/0.297229)^{1/10} - 1 .$$

The arithmetic average is 12.5%.

Clearly, the assumption of a constant inflation rate is an assumption of the grossest kind.   In view of this, it is not surprising that Section 6.3 reveals some rather large deviations of experience from the model.   The deviations in respect of 1969 and 1970 result from comparatively small experience and are possibly not significant.   However, the deviation of -28% in 1974 indicates poor performance of the model.

It is useful to note at this point that 1974 was the year in which a distinct retardation in the finalization of claims took place.   Some further changes in speed of finalization occurred in subsequent years.   This will be found to be a chronic problem in the succession of methods of analysis presented in the next few sections.   Thus the deviation of -28% is probably more attributable to this factor than to the assumption of constant inflation.

Perhaps the greatest drawback in the basic chain ladder method is that it projects the average (in some sense) past inflation rate into the future.   This has two drawbacks.   Firstly because the loading for future inflation is implicit, its magnitude is not clear.   Secondly, because the loading for future inflation is essentially an average of past inflation rates, it may be inappropriate to the future.

On the basis of the last table it seems reasonable to assume that the method has loaded for future inflation of about 12% per annum.   The outstanding claims projected by all other methods in this paper are obtained before the addition of a loading for inflation.   For the purpose of comparison, it is useful to deflate the total outstanding claims estimated in Section 6.4.   Taylor [9] found that, in rough terms each 1% p.a. future inflation added about 2.6% to outstanding claims.   Therefore, the figure of $17.705M in Section 6.4 deflates to

$$\$17.705M/(1+12\times0.026) = \$13M , \quad \text{say} ,$$

in 31/12/79 money values.

7.   METHOD 2 - INFLATION-ADJUSTED CHAIN LADDER

7.1   Method

The method is as for the basic chain ladder except that all claim payments are first brought to common dollar values according to what appears to be an appropriate claims escalation index.

The adjustment to common dollar values means that analysis is made of $C_{ij}/\lambda_{ij} = C_{ij}/\lambda_{i+j}$, instead of C_{ij} , as in (4.1).   Then all of Section 6.1 holds with C_{ij} replaced by C_{ij}/λ_{i+j} and α replaced by 1 .

7.2 Estimates of parameters

j	$M_{j,j+1}$ (a)	$M_{j\infty}$ (b)
0	3.92332	12.57329
1	1.82464	3.20476
2	1.30126	1.75638
3	1.15315	1.34975
4	1.06734	1.17049
5	1.04081	1.09665
6	1.02682	1.05365
7	1.00728	1.02612
8		1.01871

Notes : (a) *From equation (6.1.5) with* C_{ij} *replaced by* C_{ij}/λ_{i+j} .

(b) *See Section 6.2.*

7.3 Goodness-of-fit

A table of actual claim payments, and the corresponding fitted payments is given in Appendix B2.

As for the basic chain ladder (see Section 6.3), row and column totals agree as between actual and fitted payments.

A comparison of actual and fitted payments by payment year is as follows.

Payment year	Claim payments (a) (31/12/79 values)		
	actual (b)	fitted (b)	deviation (c)
	$	$	%
1969	193013	368753	- 48
1970	1960804	1543365	+ 27
1971	3280723	2986622	+ 10
1972	4506400	3974765	+ 13
1973	4214487	4556614	- 8
1974	3467526	4815047	- 28
1975	4936092	4812825	+ 3
1976	4270279	4666738	- 8
1977	5166110	4519505	+ 14
1978	4569353	4446630	+ 3
1979	4307041	4510815	- 5

Notes : (a) *For payment year shown, but only in respect of the accident years*
             included (for that payment year) in the basic data (Appendix A5).

         (b) *From Appendix B2.*

         (c) *(actual - fitted)/fitted.*

7.4 Projected outstanding claims

Outstanding claims are projected according to (6.1.12) with C_{ij} replaced by
C_{ij}/λ_{i+j} . The results are as follows.

Accident year	Estimated outstanding claims at 31/12/79 (31/12/79 values) (a)
	$M
1971	0.100
1972	0.130
1973	0.278
1974	0.403
1975	0.624
1976	1.069
1977	2.032
1978	3.025
1979	5.156
Total (1971-1979)	12.817

Note : (a) *See Note (a) to Section 6.4.*

7.5 Comment

The inflation-adjusted chain ladder allows explicitly for claims escalation, whereas
the basic chain ladder does not. A comparison of Sections 6.3 and 7.3 reveals that this
produces some small improvement in the fidelity of the model to the data. However, the
important deviation of -28% remarked upon in Section 6.5, retains precisely that value under
the inflation-adjusted method. As foreshadowed in Section 6.5, this suggests that the
deviation is due to the change in speed of finalization which occurred in 1974.

Finally, it may be noted that there is very little difference between the above
estimate of total outstanding claims (in 31/12/79 values) and the corresponding figure
obtained in Section 6.5 on the basis of the basic chain ladder.

8. METHOD 3 - PAYMENTS PER CLAIM INCURRED

8.1 Method

This is a version of the *payments per unit of risk method* originated by Cumpston [1].
In this method, (4.1) is rewritten :

$$E[C_{ij}/n_i\lambda_{i+j}] = \mu_i\rho_{ij} \ . \tag{8.1.1}$$

It is then assumed that $\mu_i\rho_{ij}$ is independent of i , and so (8.1.1) can be written :

$$E[C_{ij}/n_i\lambda_{i+j}] = \nu_j \ . \tag{8.1.2}$$

The parameters ν_j are estimated by an averaging process. Now

$$\frac{E[\sum_{i \in S} C_{ij}/\lambda_{i+j}]}{\sum_{i \in S} n_i} = \nu_j \ ,$$

when S has the same meaning as in Section 6.1. It follows, therefore, that

$$\hat{\nu}_j = \frac{\sum_{i \in S} C_{ij}/\lambda_{i+j}}{\sum_{i \in S} n_i} \tag{8.1.3}$$

is an unbiased estimator of ν_j .

It follows that an unbiased estimate of $E[C_{ij}]$ is $n_i\hat{\nu}_j\lambda_{i+j}$. Then expected outstanding claims for year of origin i are

$$E[A_{i\infty}] - E[A_{if_i}] = n_i \sum_{j=f_i+1}^{\infty} \hat{\nu}_j\lambda_{i+j} \ . \tag{8.1.4}$$

8.2 Estimates of parameters

j	$\hat{\nu}_j$ (a)
0	604
1	1773
2	2006
3	1342
4	891
5	448
6	297
7	213
8	61
9 to ∞	150 (b)
Total	7785

Notes : (a) From (8.1.3).

         (b) This is an estimate of $\sum\limits_{j=9}^{\infty} \nu_j$.

              The value has been chosen so as to produce, in common with other
              methods, a provision of approximately $0.1M in respect of development
              years 9 and 10 of accident year 1971.

              See also Note (b) to Section 6.2.

8.3 Goodness-of-fit

A table of actual claim payments and the corresponding fitted payments is given in
Appendix B3.

As for the two chain ladder methods (see Sections 6.3 and 7.3), column totals agree
as between actual and fitted payments.   Row totals are no longer constrained to agree.

A comparison of actual and fitted payments by :

(i)   accident year; and
(ii) payment year;

is as follows .

Year	Claim payments (a) (31/12/79 values) in					
	accident year shown			payment year shown		
	actual (b)	fitted (b)	deviation (c)	actual (b)	fitted (b)	deviation (c)
	$	$	%	$	$	%
1969	4551295	3961202	+ 15	193013	315892	- 39
1970	5743889	4870082	+ 18	1960804	1315651	+ 49
1971	5345859	5120024	+ 4	3287023	2597481	+ 26
1972	4995827	5097302	- 2	4506400	3596764	+ 25
1973	5175219	5955049	- 13	4214487	4366820	- 3
1974	4166594	4725816	- 12	3467526	4902880	- 29
1975	3662877	3394008	+ 8	4936092	5067720	- 3
1976	3056471	3108675	- 2	4270279	4870074	- 12
1977	2686208	2726226	- 1	5166110	4625357	+ 12
1978	1371944	1671031	- 18	4569353	4607502	- 1
1979	445545	448772	- 1	4307041	4812046	- 10

Notes : (a) See Section 7.3.

         (b) From Appendix B3.

         (c) See Section 7.3.

8.4 Projected outstanding claims

Outstanding claims are projected according to (8.1.4). The results are as follows.

Accident year	Estimated outstanding claims at 31/12/79 (31/12/79 values) (a)
	$M
1971	0.109
1972	0.142
1973	0.343
1974	0.482
1975	0.600
1976	1.119
1977	2.116
1978	3.802
1979	5.335
Total (1971-1979)	14.048

Note : *(a) See Note (a) to Section 6.4.*

8.5 Comment

In common with the preceding methods, the payments per claim incurred method makes no use of information in respect of numbers of claims finalized. It is not surprising, therefore, that it encounters difficulties when changes in speed of finalization have taken place, as is the case here.

The method is usually associated with the *direct future payments method of projection* (Sawkins, [7], [8]), as it is in Sections 8.1 and 8.4. This contrasts with the *ratio method of projection* used in Sections 6 and 7. This will result in the payments per claim incurred projection responding less rapidly to any *generation effects,* i.e. changes in average claim cost (in real terms) with increasing accident year.

The provision is of a similar magnitude to that derived by means of the inflation-adjusted chain ladder method (Section 7). This is certainly so in total, and is broadly so in respect of individual accident years.

The discrepancy between actual and fitted payments in respect of payment year 1974, which arose in the previously considered methods and was commented on in Sections 6.5 and 7.5, is still present, and of virtually unchanged magnitude.

In addition, a poor fit is obtained in the early payment years and accident years.

9. METHOD 4 - PAYMENTS PER CLAIM FINALIZED (NONCUMULATIVE)

9.1 Method

Consider the basic model (3.1) simplified according to (3.3) :

$$E[C_{ij}] = n_i \mu_i \rho_{ij} \lambda_{j+j} \qquad (9.1.1)$$

It is now assumed that average claim size is constant from one accident year to the next :

$$\mu_i = \mu , \quad \text{independent of } i . \qquad (9.1.2)$$

It is further assumed that, in expectation, the inflation-adjusted claim payments for a particular development year are proportional to the number of claims finalized in that development year (independent of accident year),

$$\text{i.e. } E[C_{ij}/\lambda_{i+j}|N_{ij}] = p_j N_{ij} , \qquad (9.1.3)$$

for some constants p_j independent of i .

By (9.1.1) to (9.1.3),

$$E[n_i \mu \rho_{ij}|N_{ij}] = p_j N_{ij} \qquad (9.1.4)$$

Therefore, for large samples,

$$E[C_{ij}/N_{ij}\lambda_{i+j}] = p_j , \qquad (9.1.5)$$

asymptotically.

If the notation :

$$P_{ij} = C_{ij}/N_{ij}\lambda_{i+j} \qquad (9.1.6)$$

is adopted, i.e. P_{ij} denotes the *inflation-adjusted payments per claim finalized* in accident year i , development year j , then (9.1.5) becomes :

$$E[P_{ij}] = p_j . \qquad (9.1.7)$$

The parameter p_j is estimated by

$$\hat{p}_j = \sum_{i \in S} n_{ij} P_{ij} / \sum_{i \in S} n_{ij} = \sum_{i \in S} (c_{ij}/\lambda_{i+j}) / \sum_{i \in S} n_{ij} , \qquad (9.1.8)$$

where p_{ij} is the observed value of the random variable P_{ij} and S has the same meaning as in Section 6.1. Thus $\hat{p}_j$ is a weighted average of the unbiased estimators P_{ij} and so is unbiased.

This method is discussed by Sawkins [7], [8] and Taylor [9].

In the projection of outstanding claims, estimates n^*_{ij} of future values of N_{ij} are obtained by some means. The determination of the n^*_{ij} is not the point of this paper and will not detain us. Then by (9.1.4), future values of C_{ij} are estimated by

$$\hat{p}_j n^*_{ij} \lambda_{i+j} . \tag{9.1.9}$$

Finally, expected outstanding claims for year of origin i are

$$E[A_{i\infty}] - E[A_{if_i}] = \sum_{j=f_i+1}^{\infty} \hat{p}_j n_{ij} \lambda_{i+j} . \tag{9.1.10}$$

9.2 Estimates of parameters

Development year j	Total number of claims finalized in development year (a)	Total inflation-adjusted amount of claim payments in development year (b)	Inflation-adjusted payments per claim finalized (c)
		$	$
0	433	4296653	9923
1	1373	11298855	8229
2	1351	11376746	8421
3	1006	6774310	6734
4	525	4013353	7644
5	207	1788079	8638
6	58	986650	17011
7	22	535783	24354
8	14	113039	8074
9			15000 (d)
10			15000 (d)

Notes : (a) *Denominator of (9.1.8). Obtained from Appendix A2.*

(b) *Numerator of (9.1.8). Obtained from Appendix A5.*

(c) *From (9.1.8).*

(d) *Assumed in order to produce payments totalling $0.1M approximately in development years 9 and 10 of accident year 1971 (see Note (b) to Section 6.2).*

9.3 Goodness-of-fit

A table of actual claim payments and the corresponding fitted payments is given in Appendix B4.

The method of fitting payments (see (9.1.8) and (9.1.9)) ensures that column totals agree as between actual and fitted payments.

A comparison of actual and fitted payments by :

(i)   accident year; and

(ii) payment year;

is as follows.

Year	Claim payments (a) (31/12/79 values) in					
	accident year shown			payment year shown		
	actual (b)	fitted (b)	deviation (c)	actual (b)	fitted (b)	deviation (c)
	$	$	%	$	$	%
1969	4551295	4479672	+ 2	193013	982377	- 80
1970	5743889	5601187	+ 3	1960804	1723724	+ 14
1971	5345859	5581914	- 4	3280723	2967961	+ 11
1972	4995827	5296908	- 6	4506400	4158657	+ 8
1973	5175219	6216824	- 17	4214487	4389417	- 4
1974	4166594	4890615	- 15	3467526	2817794	+ 23
1975	3662877	3288777	+ 11	4936092	5959661	- 17
1976	3056471	2523145	+ 21	4270279	6403074	- 33
1977	2686208	1913873	+ 40	5166110	4547633	+ 14
1978	1371944	1092274	+ 26	4569353	3636943	+ 26
1979	445545	297690	+ 49	4307041	3595638	+ 20

Notes : (a) *See Section 7.3.*

(b) *From Appendix B3.*

(c) *See Section 7.3.*

9.4 Projected outstanding claims

Outstanding claims are projected according to (9.1.9).   The results are as follows :

Accident year	Estimated outstanding claims at 31/12/79 (31/12/79 values) (a)
	$M
1971	0.105
1972	0.178
1973	0.444
1974	0.987
1975	1.328
1976	2.612
1977	4.200
1978	6.072
1979	7.288
Total (1971-1979)	23.214

Note : (a) *See Note (a) to Section 6.4.*

9.5 Comment

Of the four methods considered hitherto, the current one is the first to attempt to take account of speed of finalization. Although the elaboration of the model - the assumption that, in any combination of year of origin and development year, claim payments will be proportional to number of claims finalized - appears a sensible one, it is abundantly clear that it has failed in its aim.

The large deviation of experience from the model in payment year 1974, observed in connection with other methods, has been reduced only slightly. This deviation was suspected to result from changing speed of finalization; and yet the allowance for this factor in the current method has not achieved a markedly better fit.

Furthermore, the small reduction in the 1974 deviation is matched by the introduction of quite large deviations in subsequent payment years.

Even further, the model appears to go badly awry in the recent accident years.

Comparisons between the models are not easy because different numbers of degrees of freedom are involved. Allowance for this is difficult without a great deal of empirical analysis, as variances of the different items of data are unknown. However, it appears that the current model, the first to attempt to allow for speed of finalization, may well provide the *worst* fit out of all four models so far considered.

This is a salutary indication of the care required in the introduction of a new factor into the model.

10. METHOD 5 - PAYMENTS PER CLAIM FINALIZED (CUMULATIVE)

10.1 Method

The main difference in principle between this and the previous method is that the assumption (9.1.2) is now dropped. That is, average claim size is assumed possibly to vary from one year of origin to another. In these circumstances the assumption described by (9.1.3) is not reasonable. In other words, if average claim size is assumed to be varying over year of origin, it is not reasonable to assume that payments (inflation-adjusted) per claim finalized in development year j will be independent of year of origin.

If variation of μ_i with i is retained, then (9.1.3) is replaced by :

$$E[C_{ij}/\lambda_{i+j}|N_{ij}] = \mu_i q_j N_{ij} , \tag{10.1.1}$$

which can be written in the form :

$$E[C_{ij}/\lambda_{i+j}|N_{ij}] = (\mu_i/\mu) \times (\mu q_j)N_{ij} , \tag{10.1.2}$$

where μ is an average claim size over all years of origin (just what type of average is not important; all that is required is some *standard* claim size). Note that, if assumption (9.1.2) is now applied, μq_j becomes equal to p_j . Thus, in the current model, it is possible to identify μq_j as the expected payments per claim finalized in development year j *when the standard claim size applies*; and μ_i/μ as a factor to adjust the expected payments per claim finalized up or down according as average claim size for the year of origin in question exceeds or falls short of the standard.

Let

$$B_{ij} = \sum_{k=0}^{j} C_{ik}/\lambda_{i+k}$$

$$= \text{inflation-adjusted version of } A_{ij} .$$

Then, by (10.1.1),

$$E[B_{ij}|\{N_{ij}\}] = \mu_i \sum_{k=0}^{j} q_k N_{ik} , \tag{10.1.3}$$

and, asymptotically for large samples,

$$E\left[B_{ij}/\sum_{k=0}^{j} N_{ik}|\{N_{ij}\}\right] = \mu_i \sum_{k=0}^{j} q_k N_{ik}/\sum_{k=0}^{j} N_{ik} . \tag{10.1.4}$$

Now write

$$R_{j,j+1} = \left[B_{i,j+1}/\sum_{k=0}^{j+1} N_{ik}\right]/\left[B_{ij}/\sum_{k=0}^{j} N_{ik}\right] \tag{10.1.5}$$

By (10.1.4) and (10.1.5),

$$E[R_{j,j+1}|\{N_{ij}\}] = \frac{\displaystyle\sum_{i\in S}\sum_{k=0}^{j+1} q_k N_{ik}}{\displaystyle\sum_{i\in S}\sum_{k=0}^{j+1} N_{ik}} \bigg/ \frac{\displaystyle\sum_{i\in S}\sum_{k=0}^{j} q_k N_{ik}}{\displaystyle\sum_{i\in S}\sum_{k=0}^{j} N_{ik}} , \qquad (10.1.6)$$

asymptotically.

Clearly, the expected value of $R_{j,j+1}$, conditional on the N_{ij} , depends on the values of $N_{ik}/\sum_{k=0}^{j} N_{ik}$, i.e. on speed of finalization. However, if it is assumed that variations in speed of finalization will be sufficiently small that the ratio of weighted averages will not vary much, then (10.1.6) can be written approximately as

$$E[R_{j,j+1}|\{N_{ij}\}] = \pi_j . \qquad (10.1.7)$$

The parameter π_j is the factor by which average claim payment per claim finalized *up to date* changes between the end of development year j and the end of development year $j+1$.

By (10.1.6) and (.10.1.7), the right side of (10.1.5) is an asymptotically unbiased estimator $\hat{\pi}_j$ of $E[R_{j,j+1}|\{N_{ij}\}]$.

Now write,

$$R_{jk} = R_{j,j+1} R_{j+1,j+2} \cdots R_{k-1,k} , \qquad k > j ,$$

whence, asymptotically,

$$E[R_{jk}|\{N_{ij}\}] = \pi_j \pi_{j+1} \cdots \pi_{k-1} . \qquad (10.1.8)$$

Thus $\hat{\pi}_j \hat{\pi}_{j+1} \cdots \hat{\pi}_{k-1}$ is an asymptotically unbiased estimate of $E[R_{jk}|\{N_{ij}\}]$.

The projection proceeds as follows. As in Section (9.1), it is necessary to predict future numbers n_{ij}^{*} of finalizations.

By (10.1.5),

$$B_{i,j+1} = R_{j,j+1} \left(\sum_{k=0}^{j+1} N_{ik}\right) \left[B_{ij}/\sum_{k=0}^{j} N_{ik}\right] ,$$

and, by extension,

$$B_{ik} = R_{jk} \left(\sum_{\alpha=0}^{k} N_{i\alpha}\right) \left[B_{ij}/\sum_{\alpha=0}^{j} N_{i\alpha}\right] ,$$

whence

$$E[B_{ik}|\{N_{ij}\}] = \hat{\pi}_{jk} \left(\sum_{\alpha=0}^{k} N_{i\alpha} \right) E\left[B_{ij} / \sum_{\alpha=0}^{j} N_{i\alpha} \right] ,$$

asymptotically, where $\hat{\pi}_{jk}$ denotes $\hat{\pi}_j \hat{\pi}_{j+1} \cdots \hat{\pi}_{k-1}$. Thus an asymptotically unbiased estimate of $E[B_{ik}\{N_{ij}\}]$ is

$$\hat{\pi}_{jk} \left(\sum_{\alpha=0}^{k} N_{i\alpha} \right) \left[b_{ij} / \sum_{\alpha=0}^{j} N_{i\alpha} \right] , \qquad (10.1.9)$$

where b_{ij} is the observed value of B_{ij}. The choice $j = f_i$ bases the estimator on the maximum amount of information, in which case the estimate of $E[B_{ik}|\{N_{ij}\}]$ is :

$$\hat{\pi}_{f_i k} \left(\sum_{\alpha=0}^{k} N_{i\alpha} \right) \left[b_{if_i} / \sum_{\alpha=0}^{f_i} N_{i\alpha} \right] . \qquad (10.1.10)$$

The expression (10.1.10) may be differenced with respect to k to produce estimates of $E[C_{ij}/\lambda_{i+j}|\{N_{ij}\}]$. Also, the amount of outstanding claims, in current dollar values, in respect of year of origin i is

$$\hat{\pi}_{f_i \infty} N_i \left[b_{if_i} / \sum_{\alpha=0}^{f_i} N_{i\alpha} \right] - b_{if_i} . \qquad (10.1.11)$$

To the author's knowledge, the method has not previously appeared in the literature in precisely this form. It is however, a ratio version of Sawkins' [7], [8]· cumulative payments per claim finalized method.

10.2 Estimates of parameters

Development year j	Development factor	
	$\hat{\pi}_j$ (a)	$\hat{\pi}_{j\infty}$ (b)
0	0.893	0.904
1	1.019	1.012
2	0.951	0.993
3	1.000	1.045
4	1.008	1.045
5	1.022	1.036
6	1.016	1.014
7	0.998	0.998
8		1.000 (c)

Notes : *(a) See (10.1.5).*

        (b) Calculated as $\hat{\pi}_j \hat{\pi}_{j+1} \cdots$.

        (c) It is assumed that $\hat{\pi}_{j\infty} = 1$ *for* $j \geq 8$.

10.3 Goodness-of-fit

A table of actual claim payments and the corresponding fitted payments is given in Appendix B5.

The method of fitting (see Note (b) to Appendix B5) ensures that row totals agree as between actual and fitted payments and that column totals effectively agree.

A comparison of actual and fitted payments by payment year is as follows.

Payment year	Claim payments (a) (31/12/79 values)		
	actual (b)	fitted (b)	deviation (c)
	$	$	%
1969	193013	956674	- 80
1970	1960804	1721796	- 14
1971	3280723	3211994	+ 2
1972	4506400	3792835	+ 19
1973	4214487	4114127	+ 2
1974	3467526	2536449	+ 37
1975	4936092	5403325	- 9
1976	4270279	5868200	- 27
1977	5166110	4650135	+ 11
1978	4569353	4415397	+ 3
1979	4307041	4512893	- 5

Notes : *(a) See Section 7.3.*

        (b) From Appendix B5.

        (c) See Section 7.3.

10.4 Projected outstanding claims

Outstanding claims are projected according to (10.1.11). The results are as follows.

Accident year	Payments per claim finalized up to 31/12/79 (a) (b)	Ultimate development factor (c)	Ultimate average claim cost (a) (d)	Estimated outstanding claims at 31/12/79 (a) (e)
	$		$	
1971	7991	1.000	7991	0.056
1972	7604	0.998	7589	0.111
1973	6661	1.014	6754	0.289
1974	6819	1.036	7064	0.560
1975	8742	1.045	9135	1.023
1976	9642	1.045	10075	2.415
1977	12046	0.993	11962	4.754
1978	10718	1.012	10847	6.253
1979	14852	0.904	13426	9.530
Total (1971-1979)				24.991

Notes : (a) *In 31/12/79 values.*

(b) *Calculated from Appendices A2 and A5.*

(c) *Value of* $\hat{\pi}_{f_i \infty}$.

(d) *Product of preceding two columns.*

(e) *From (10.1.11).*

10.5 Comment

The method has the important advantage over the preceding one that it allows for "generation effects" in the form of an average claim size which varies with year of origin. However, it will be observed that, in the development of (10.1.7), the model involves itself in something of a contradiction. Although one of its fundamental aims is to allow for variation in speed of finalization, it is necessary in making use of (10.1.6) to assume that the effects of such variations are not too great. Thus, in precisely the circumstances in which the inclusion of speed of finalization in the model would be most useful, the reliability of this method becomes doubtful.

Perhaps most significantly, the fit of the model to the data obtained by this method is not appreciably better than by the previous method (Section 9). The problem year 1974 again throws up a large deviation (in fact, the largest observed in any of the methods). In common with the previous method, a large deviation also occurs in payment year 1976.

A better fit is obtained for accident years 1977 to 1979. This results from the extra constraints ensuring equality between actual and fitted total payments for each accident year.

It is this set of constraints which builds in the allowance for the generation effects mentioned above. The generation effects appear in the form of sharply increased average claim sizes in the recent accident years. It appears, however (see Section 11.5), that these effects may have been over-estimated by the present method. Certainly, the jump in average claim size between accident years 1978 and 1979, as produced by the present method, appears large.

11. METHOD 6 - REDUCED REID METHOD

11.1 Method

This method is described by Taylor [9]. It is similar to the preceding method in most respects. It does, however, rest on the premise that, for a particular year of origin, the cumulative payments per claim finalized up to any given point of development is not a function of the period of that development measured in real time (as in Section 10) but of the proportion of that year's claims which have been finalized (called *operational time*). This type of assumption was introduced by Reid [5].

After this explanation, the method can be seen as a rationalization of the preceding one. Consider (10.1.6). If the ratios $N_{ik} / \sum_{k=0}^{j} N_{ik}$ are independent of i, then the right side of (10.1.6) is independent of i. Then the assumption introduced just after (10.1.6) is no longer necessary, and (10.1.7) is no longer just an approximation.

Now the way to ensure that the $N_{ik} / \sum_{k=0}^{j} N_{ik}$ are independent of i is to *define them that way*. But all this means is that one considers cumulative claim payments per claim finalized at various operational times instead of at various real times (ends of development years).

It follows from the above that all of the Section 10.1 goes through for the Reduced Reid method except that throughout "at end of development year j" is translated into "at operational time j".

In practical terms this means that interpolations are required to produce values of

$$ B_{ij} / \sum_{k=0}^{j} N_{ik} \quad \text{and} \quad \hat{\pi}_{j,1} $$

at the required values of operational time j. In the present application, each of these functions is assumed to be piecewise linear, although clearly other forms of interpolation are possible.

Complete details of the application of the method to the data of Appendix A are given by Taylor [9].

11.2 Estimates of parameters

Operational time, t_j	Development factor	
	$\hat{\pi}_{t_j, t_{j+1}}$ (a)	$\hat{\pi}_{t_j, 1}$ (b)
0.04	0.937	0.611
0.07	0.753	0.652
0.18	0.889	0.866
0.36	0.948	0.974
0.58	0.977	1.027
0.70	0.983	1.051
0.82	1.004	1.069
0.91	1.021	1.065
0.960	1.034	1.043
0.976	1.009	1.009
0.990	(c)	1.000 (d)

Notes : (a) *Calculated from (10.1.5) with real time j replaced by operational time t_j .*

(b) *Calculated by compounding the factors of the preceding column.*

(c) *Not calculated.*

(d) *Assumed.*

11.3 Goodness-of-fit

A table of actual claim payments, and the corresponding fitted payments is given in Appendix B6.

The process by which fitted claim payments have been obtained causes indentity of total actual and fitted claim payments in each year of origin (see Note (b) to Appendix B6).

A comparison of actual and fitted payments by :

(i) development year; and

(ii) payment year;

is as follows.

Development year	Claim payments (in 31/12/79 values)			Payment year	Claim payments (a) (in 31/12/79 values)		
	actual(b)	fitted(b)	deviation (c)		actual(b)	fitted(b)	deviation (c)
	$	$	%		$	$	%
0	4296653	5569154	- 23	1969	193013	990114	- 81
1	11298855	11363398	- 1	1970	1960804	1846068	+ 6
2	11376746	9922220	+ 15	1971	3280723	2938921	+ 12
3	6774310	6951550	- 3	1972	4506400	4055217	+ 11
4	4013353	3787454	+ 6	1973	4214487	4127440	+ 2
5	1788079	2121677	- 16	1974	3467526	2659053	+ 30
6	986650	873608	+ 13	1975	4936092	5342481	- 8
7	535783	421989	+ 27	1976	4270279	5927761	- 28
8	113039	183428	- 38	1977	5166110	4959784	+ 4
				1978	4569353	4291222	+ 6
				1979	4307041	4056417	+ 6

Notes : (a) *See Section 7.3.*

(b) *From Appendix B6.*

(c) *See Section 7.3.*

11.4 Projected outstanding claims

Outstanding claims are projected according to (10.1.11) but with real times replaced by the corresponding operational times.   Full details are given in Taylor [9].

Accident year	Payments per claim finalized up to 31/12/79 (a) (b)	Ultimate development factor (c)	Ultimate average claim cost (a) (d)	Estimated outstanding claims at 31/12/79 (a)(e)
	$		$	$M
1971	7991	1.000	7991	0.056
1972	7604	1.009	7672	0.168
1973	6661	1.043	6947	0.445
1974	6819	1.065	7262	0.692
1975	8742	1.069	9345	1.131
1976	9642	1.027	9902	2.320
1977	12046	0.974	11733	4.612
1978	10718	0.866	9282	5.153
1979	14852	0.611	9075	6.297
Total (1971-1979)				20.874

Notes :   (a)   *In 31/12/79 values.*

         (b)   *Calculated from Appendices A2 and A5.*

         (c) $\hat{\pi}_{f_i 1}$, *where* f_i *is the operational time attained by accident year* i *at 31/12/79.*

         (d)   *Product of preceding two columns.*

         (e)   *Calculated from (10.1.11) with real times replaced by corresponding operational times.*

11.5 Comment

In general terms, the Reduced Reid method produces a very similar fit to that produced by the cumulative payments per claim finalized method (compare Sections 10.3, 11.3).   The difficulty with payment year 1974 is less for the Reduced Reid method but still very significant.

However, despite this similarity in fit, a difference of about \$4M exists between the provisions produced by the two methods (compare Sections 10.4, 11.4).   This difference relates almost entirely to the two most recent accident years, 1978 and 1979.   Because of the similarity in fit of the two models to the data, it is difficult to decide, on the basis of examination of the two methods alone, which is likely to be the more realistic.   However, if Section 12 is taken into account, then it appears that the Reduced Reid method has under-estimated in respect of these two accident years and the cumulative payments per claim finalized method overestimated.   The more realistic result appears to be about mid-way between the two.   See also the remarks in Section 13.

12. METHOD 7 - SEE-SAW METHOD

12.1 Method

This is a new method.

The basic data, presented in Appendix C, are derived from Appendix A.   If one examines the payments per claim finalized (in a given development year) in Appendix C1 in conjunction with the speeds of finalization in Appendix C2, there appears a tendency toward inverse correlation between speed of finalization and payments per claim finalized.   This is not easy to detect by inspection because of the confusing effect of operational time, which develops differently for different years of origin.   Nevertheless, it does appear that, on the whole, as speed of finalization go up, payments per claim finalized come down.   Hence the name of the method.

This inverse correlation stands to reason.   Claim payments in a year consist of :

(i)   payments in respect of claims finalized in the year;

(ii)   *partial payments*, i.e. payments in respect of claims which still remained open with an outstanding balance at the end of the year.

Thus in the ratio of payments per claim finalized there is not a one-to-one correspondence between numerator and denominator. The numerator consists of the above two components, of which the first is in one-to-one correspondence with the denominator, but the second is not. It may be reasonable to assume that the first component of the numerator will be proportional to the denominator. However, the second usually will not be. It appears in practice that partial payments are often rather insensitive to changes in speed of finalization, causing a negative correlation between speed of finalization and (total) payments per claim finalized.

The observation of an inverse relationship between speed of finalization and payments per claim finalized represents a refinement of the model used in Section 3. We now have

$$E[C_{ij}/\lambda_{i+j}|N_{ij},F_{ij}] = g(t_j,N_{ij},F_{ij}) \ , \tag{12.1.1}$$

where

F_{ij} = speed of finalization in year of origin i , development year j

$\cdot$ = proportion of year of origin i claims incurred finalized in development year j ;

t_j = average operational time during development year j ;

g is some function which decreases with increasing F_{ij} .

In this application, we choose g to be of the following form :

$$g(t_j,N_{ij},F_{ij}) = N_{ij}[\alpha(t_j) + \beta(t_j)t_j + \gamma(t_j)F_{ij}] \tag{12.1.2}$$

where α , β , γ are step functions with steps over ranges determined by the user of the method, and where $\alpha(t_j) + \beta(t_j)t_j$ is continuous in t_j .

By (12.1.1) and (12.1.2),

$$E[C_{ij}/\lambda_{i+j}N_{ij}|N_{ij},F_{ij}] = \alpha(t_j) + \beta(t_j)t_j + \gamma(t_j)F_{ij} \ , \tag{12.1.3}$$

asymptotically. On account of the continuity requirement mentioned just after (12.1.2), it is convenient to rewrite (12.1.3) as

$$E[P_{ij}|N_{ij},F_{ij}] = E[C_{ij}/\lambda_{i+j}N_{ij}|N_{ij},F_{ij}]$$

$$= \alpha + \sum_k \beta_k t_j^{(k)} + \sum_k \gamma_k F_{ij}^{((k))} \ , \tag{12.1.4}$$

where

$t_j^{(k)} = t_j$, provided that t_j lies in the k-th range of operational time;

= the lower bound of the k-th range if t_j is less than this;

= the upper bound of the k-th range of t_j is greater than this;

$F_{ij}^{((k))} = F_{ij}$, provided that t_j lies in the k-th range of operational time;

= 0 , otherwise.

α = constant term.

By (12.1.4), the random variable P_{ij} may be regressed on the $t_j^{(k)}$, $F_{ij}^{((k))}$ and so the α , β_k , γ_k estimated.   Once this has been done it is a simple matter to estimate $E[P_{ij}|N_{ij},F_{ij}]$ for any given values of N_{ij} , F_{ij} .   For the projection, future values n_{ij}^* , f_{ij}^* of N_{ij} , F_{ij} are predicted and so estimates $\hat{p}_{ij}$ of future $E[P_{ij}|N_{ij},F_{ij}]$ obtained.

Outstanding claims are then estimated for year of origin i as :

$$E[A_{i\infty}] - E[A_{if_i}] = \sum_{j=f_i+1}^{\infty} n_{ij}^* \hat{p}_{ij} \lambda_{i+j} .$$ (12.1.5)

12.2 Estimates of parameters

Range of operational time k	from operational time	to operational time	β_k (a)	γ_k (a)	α (a)
			$	$	$
1	0	0.15	− 2191	−33950	
2	0.15	0.35	+10410	−22090	
3	0.35	0.55	− 1194	−30380	
4	0.55	0.75	−16720	−30870	−188100
5	0.75	0.85	− 45.09	−35260	
6	0.85	0.95	+14660	−38010	
7	0.95	1.00	+208100	−145200	

Notes : (a) *Estimated by means of a weighted regression in accordance with (12.1.4).*

The weight associated with P_{ij} was N_{ij} .   The regression was carried out by means of the GLIM package.   The error terms were assumed to be normally distributed.

12.3 Goodness-of-fit

A table of actual claim payments and the corresponding fitted payments is given in Appendix B7.

A comparison of actual and fitted payments by :

(i)   accident year;

(ii) development year;

(iii) payment year;

is as follows.

Acci-dent year	Claim payment (in 31/12/79 values)			Deve-lop-ment year	Claim payments (in 31/12/79 values)			Pay-ment year	Claim payments (a) (in 31/12/79 values)		
	actual (b)	fitted (b)	devi-ation (c)		actual (b)	fitted (b)	devi-ation (c)		actual (b)	fitted (b)	devi-ation (c)
	$	$	%		$	$	%		$	$	%
1969	4551295	4411005	+ 3	0	4296653	4734478	- 9	1969	193013	720918	- 73
1970	5743889	5403078	+ 6	1	11298855	10945964	+ 3	1970	1960804	1934714	+ 1
1971	5345859	5591369	- 4	2	11376746	10712276	+ 6	1971	3280723	2933401	+ 12
1972	4995827	5404176	- 8	3	6774310	7116831	- 5	1972	4506400	3848469	+ 17
1973	5175219	5324586	- 3	4	4013353	4182092	- 4	1973	4214487	4516072	- 7
1974	4166594	4408204	- 6	5	1788079	2190445	- 19	1974	3467526	3457015	+ 0
1975	3662877	3972012	- 9	6	986650	801170	+ 23	1975	4936092	5249940	- 6
1976	3056471	3025640	+ 1	7	535783	382910	+ 40	1976	4270279	4755528	- 10
1977	2686208	2227982	+ 21	8	113039	274130	- 59	1977	5166110	4864408	+ 6
1978	1371944	1201360	+ 14					1978	4569353	4404646	+ 4
1979	445545	376200	+ 18					1979	4307041	4659490	- 8

Notes : (a) See Section 7.3.

(b) From Appendix B7.

(c) See Section 7.3.

12.4 Projected outstanding claims

Outstanding claims are projected according to (12.1.5). The predictions of n^*_{ij} are the same as in Taylor [9] and are repeated in Appendix C3. The values $\hat{p}_{ij}$ fitted to the future P_{ij} are also given in Appendix C3.

The results of the projection are as follows.

Accident year	Outstanding claims at 31/12/79 (in 31/12/79 values)
	$M
1971	0.156
1972	0.322
1973	0.572
1974	0.810
1975	1.110
1976	2.262
1977	4.194
1978	6.108
1979	7.269
Total (1971-1979)	22.803

12.5 Comment

By far the most important point is that the see-saw method has significantly reduced the deviation between experience and the model in payment year 1974 (coincidentally, it has eliminated the deviation almost completely).

This has not been done at the expense of introduction of unreasonably large deviations elsewhere.   The deviation in payment year 1972 appears dubious, but it is not dramatic.   The only other areas in which large deviations occur are the recent accident years and the high order development years, in each of which cases the experience is relatively small.   In these cases it is difficult to decide whether the deviations are of a random nature or indicative of generations effects (see Section 10.5).   Their size and persistent positivity may cause one to be a little wary of the latter.   But the evidence is hardly conclusive.

On the whole, it is fair to say that the see-saw method is the only one of the seven considered in this paper which provides a reasonable fit of the relevant model to the data.

13. A "CONTROL RESULT"

The range of results obtained in Sections 6 to 12 is summarised in the following table.

Method	Provision for outstanding claims (a) at 31/12/79 (31/12/79 values)
	$M
Chain ladder	
- basic	c. 13
- inflation adjusted	12.8
Payments per claim incurred	14.0
Payments per claim finalized	
- noncumulative	23.2
- cumulative	25.0
Reduced Reid	20.9
See-saw	22.8

Note : (a) Accident years 1971 and later.

The results fall immediately into two groups. The three methods which take no account of numbers of claims finalized form the low group; the methods which do take account of these numbers form the high group. It is necessary to choose between the two groups.

The fact that the see-saw method is the only one to produce a reasonable fit of model to experience, and that its result lies in the middle of the high group, suggests that the high group is likely to be the appropriate choice.

This suggestion is in fact confirmed by the existence of a "control result". The control is based on case estimates (or physical estimates) of individual outstanding claims. History has shown these estimates, in the company concerned, to be of relatively high quality. Certainly, they appear to achieve stability within two years of the accident date. The manner in which they are revised in these two years is, on the basis of history, reasonably predictable.

Use of these physical estimates, together with what appear to be appropriate adjustments in respect of the early development years, has led to an alternative estimate of outstanding claims. For accident years 1971 and later, that estimate was \$21.7M. This is about mid-way between the results of the Reduced Reid and See-saw methods, the two most sophisticated of the above seven methods.

Of course, the "control" is subject to its own qualifications, just like the results obtained by the other methods. And so, it is not necessarily to be accepted in preference to some of the results in the above table.

A reasonable assessment of all the results would, I believe, run as follows :

(i) The first three results can be discarded.

(ii) The fourth method can probably be discarded (see, however, the remarks of Section 14) on account of its poor fit of model to experience. The method is useful, however, in that the poorness of the fit (Section 9.3) contains a hint that claim sizes have been increasing more rapidly than inflation in the recent accident years.

(iii) Allowance for these "generation effects" is made in the next two methods. These two methods unfortunately give widely differing results of \$25.0M and \$20.9M.

(iv) The final method gives a result about mid-way between the preceding two. It has the drawback, however, that it does *not* make allowance for generation effects. There is some suggestion in Section 12.3 that failure to allow for them may be causing underestimation in respect of the last three accident years. This then would suggest a provision greater than the \$22.8M estimated by the See-saw method.

(v) On the other hand, the result obtained on the basis of the case estimates if *lower* than \$22.8M. If the suspected generation effects were in fact present, one would expect them to be manifesting themselves in such a way as to be affecting the case estimates - at least in respect of 1977, the oldest of the three accident years under suspicion.

(vi) A reasonable compromise between the viewpoints expressed in (iv) and (v) is
      probably to adopt the See-saw result of $22.8M.

14. CONCLUSION

As has been pointed out on previous occasions, e.g. Taylor [9, Section 2], claim
number information can occasionally be of dubious value. Essentially, this results from the
possibility of a redefinition by the administration of what constitutes a finalized claim.

This has been seen by previous writers as a reason for regarding results based on claim
numbers with caution. Unfortunately, no means of determining the extent to which this
causion is justified seems to have been suggested.

The foregoing subsections on goodness-of-fit underline my own approach to this issue.
If a sensible model involving claim numbers is being used, and if experience appears to fit
that model reasonably well, then, whatever effects have been injected into claim numbers by
management action, the model is probably reasonable for projection of outstanding claims.

The possibility of using number of claims handled (rather than finalized) was raised in
Section 2. In the case of Compulsory Third Party claims, which generated the data used in
this paper, finalization of a claim tends to be dominated by one or a few relatively large
payments. In these circumstances, claim payments in a year seem more likely to be related
to numbers of claims finalized than to numbers of claims handled, the latter category including
significant numbers of claims in respect of which no action has been taken during the year.
The use of numbers of claims handled may be more appropriate for a class of business such as
Workers Compensation insurance where claims are being paid by periodic instalments of
compensation. However this is only surmise, and I am not aware of any hard evidence one way
or the other.

The see-saw method, as fully developed in Section 12, seems appropriate for application
to Compulsory Third Party and Public Liability lines of business. Although Employers Liability
business is also of the long-tail variety, the see-saw method is not so useful in that
connection. The reason for this has to do with the manner in which *number* of outstanding
claims runs off. As is apparent from Appendix C2, in Compulsory Third Party business
significant numbers of claims are still being finalized 5 years after the accident year. This
is in contrast with Employers Liability business, where the great bulk of claims (in terms of
numbers) have been finalized by the end of development year 1. The small number of
finalizations after this makes application of the see-saw method rather trivial.

Nevertheless, the inverse correlation which forms the basis of the see-saw method is to
be observed in Employers Liability business. In view of this, I prefer to apply the see-saw
method on a non-rigorous basis to Employers Liability insurance. This is done by drawing up
tables such as those in Appendix C, but determining future values of payments per claim
finalized by inspection rather than by a formal regression. For Compulsory Third Party, and
probably Public Liability business, a full regression seems preferable.

The estimation of outstanding claims, in common with other exercises in statistical estimation, involves two problematic phases :

(i)   design of a model;

(ii) estimation of the parameters of that model.

What emerges from the preceding sections is just how important is model design relative to parameter estimation.   For example, the goodness-of-fit achieved by the inflation-adjusted chain ladder method (Section 7.3) and the cumulative payments per claim finalized method (Section 10.3) respectively are not strikingly different.   Yet, as a result of difference in model design, the latter produces an estimate of outstanding claims almost double that produced by the former.   In these circumstances, it becomes apparent that the model must be chosen with great care, and its goodness-of-fit to the experience examined.

The penalty to be paid for an incorrect choice of model is such that the importance of model design seems to outweigh by far questions of technique in estimation of parameters.

The foregoing sections also contain an implicit warning against the use of summary statistics.   It will be noticed that all of the subsections concerned with goodness-of-fit have consisted merely of tabulating total actual and model payments for given accident years, development years or payment years.   No significance tests have been carried out.   This is, of course, partly due to the fact that, as the variance of individual items of data are completely unknown, significance tests are not available.   However, there is a little more to the story than this.   Despite lack of knowledge of variances, one might be attempted to construct a statistic of the chi-squared type, i.e. $\Sigma(A-E)^2/E$.   Use of such a statistic as a criterion for assessment of goodness-of-fit of the various models might well lead to disaster.   Indeed, the failure of the earlier models of this paper to fit the experience tends to be on the subtle side, manifesting itself in most cases in only one or two isolated areas. For example, the Reduced Reid model appears to fit the experience reasonably well except for payment years 1974 and 1976.   As regards these payment years in particular, the improvement in fit produced by the see-saw method is quite outstanding.   However, these isolated areas of success or failure in fit would be swamped by the amount of other information included in a summary statistic of the type displayed above.   There seems no real alternative to a critical scrutiny of tables such as those appearing in the preceding subsections on goodness-of-fit.

Further emphasis of the importance of model design occurs in Section 9.3.   This deals with the non-cumulative payments per claim finalized method, the first method to attempt to make allowance in the model for speed of finalization.   Note that the fit of the model to the experience is significantly worse than in any of the cases preceding it, in which no allowance is made for speed of finalization.   Thus, although an extra factor has been introduced into the model, and that a factor requiring introduction, the fit has been worsened.   The implication is that the model structure according to which the additional factor has been incorporated is incorrect.

The see-saw method, on the other hand, incorporates the same additional factor, speed
of finalization, but does so in such a way that the model fits the experience (with some slight
reservations - see Section 12.5).   Application of the method to a single set of data does not,
of course, prove that the see-saw model is universally applicable.   It does, however,
indicate strongly that the model is appropriate in this case, and provides a point of
departure for the investigation of other sets of data.

One of the drawbacks of the see-saw method, mentioned in Section 12.5, ought to be
repeated.   It is that the method includes no allowance for "generation effects".   That is,
average claim size, after adjustment for inflation, is assumed to remain essentially constant
from one year of origin to another.   This may not be an appropriate assumption for the
reasons given in Taylor [9, Section 10].   It seems clear that, if such effects are to be
built into the model, the number of parameters involved in the model must increase
significantly.   The danger of overfitting then arises.   At the moment, the only solution
to this problem appears to be use of the Reduced Reid method in conjunction with the see-saw
method.

While on the subject of number of parameters, it is probably as well to anticipate and
lay to rest a possible criticism that the see-saw method already involves overfitting on
account of the multiple regression which it uses.   This view would be false.   Of the seven
methods presented in this paper, all have degrees of freedom of the same order, and some a
smaller a number of degrees of freedom than the see-saw method.   The actual numbers of degrees
of freedom are as follows.

Method	Number of degrees of freedom
	$M
Chain ladder	
- basic	46
- inflation-adjusted	46
Payments per claim incurred	54
Payments per claim finalized	
- noncumulative	54
- cumulative	46
Reduced Reid	44
See-saw	48

Note :   *There are 63 cells in the claim payment array (see Appendix A3).*
          Therefore, in each of the above cases the difference between 63
          and the number of degrees of freedom is equal to the number of
          parameters estimated.

Despite the fact that the see-saw method does not involve overfitting, the question
nevertheless arises as to whether any simplification of the model is possible.   A scrutiny
of the results of the regression involved suggests that the effect of speed of finalization is

much greater than the effect of operational time.   It may therefore be possible to rework the see-saw method on the basis of real time and retain a reasonable fit of model to experience. This has not been attempted.   However, even if such a course were successful, it is not clear that a great deal would be achieved.   One of the benefits of being forced to a multiple regression approach is that the estimation of the parameters involved needs to be formalized and carried out properly.   The distribution of error terms needs to be considered (in Section 12 both gamma and normal error terms were tried before a decision was finally made in favour of the latter); the weights to be applied to the various items of data in the regression needs to be considered.   There are some advantages in being forced to consider these questions rather than using the standard procedures of merely totalling up rows, averaging columns, and so on. If simplification of the see-saw method were to be accompanied by a reversion to these standard procedures, then such advantages would be lost.   If the see-saw model were simplified, but a proper regression analysis still carried out, then the complexity of the model *per se* would be reduced but not much work saved.

One final very interesting point emerges from a comparison of Sections 9 and 12. Section 9.3 indicates that the noncumulative payments per claim finalized method provides perhaps the worst fit of model to data out of all seven methods considered; whereas Section 12.3 indicates that the see-saw method provides the best fit.   Yet despite the yawning gulf between these two methods in terms of efficiency, comparison of Sections 9.4 and 12.4 reveals that the provisions produced by them are extremely similar.   It seems clear from the poorness of fit of the noncumulative payments per claim finalized method that this similarity would not always hold.   However, it does give rise to a question as to whether the method possesses some subtle robustness which is concealed by the poorness-of-fit displayed in Section 9.3.   An investigation of the existence or otherwise of such robustness, and of the circumstances in which the payments per claim finalized method gives satisfactory results, would be most illuminating.

APPENDIX A

DATA

A1.   Numbers of claims incurred   (n_i in the notation of Section 4)

Accident year	Number of claims incurred (a)
1969	523
1970	643
1971	676
1972	673
1973	809
1974	669
1975	513
1976	543
1977	622
1978	703
1979	743

Note : (a) *These numbers consist of numbers of claims reported to the end-1979*
             plus estimated numbers incurred but not reported (IBNR). Because
             the second of these components has been estimated, the above numbers
             are strictly subject to estimation error. However, the scope for
             error is quite small. For years other than the very latest it is
             very small indeed. Throughout this paper, estimation error in the
             N_i *has been diregarded.*

A2. Numbers of claims finalized (n_{ij} in the notation of Section 3)

Accident year	Number of claims finalized in development year								
	0	1	2	3	4	5	6	7	8
1969	99	154	112	84	37	21	7	5	2
1970	46	193	187	89	34	43	27	9	9
1971	44	191	193	78	99	49	10	2	3
1972	45	166	78	185	136	36	5	6	
1973	52	115	256	240	78	27	9		
1974	25	140	216	129	70	31			
1975	16	93	126	113	71				
1976	16	114	99	88					
1977	37	102	84						
1978	23	105							
1979	30								

A3. Claim payments (c_{ij} in the notation of Section 3)

Accident year	Claim payments in development year								
	0	1	2	3	4	5	6	7	8
	$	$	$	$	$	$	$	$	$
1969	57369	514096	418034	305728	208822	77754	82811	52596	20405
1970	122161	559595	674860	453750	156919	246291	200516	88166	20440
1971	212991	619776	561523	398485	438079	161268	120262	50451	61742
1972	168274	426536	518479	705550	378684	400212	214705	285566	
1973	201373	536473	912775	707685	716962	246320	187767		
1974	191049	612854	956866	611637	484568	260301			
1975	181845	511027	942889	798263	599887				
1976	198670	850127	1074810	502516					
1977	455912	830559	1129424						
1978	318237	963590							
1979	424327								

A4. Inflation index (λ_{i+j} in the notation of Section 3)

Calendar year	Claims inflation index (a)
1969	0.297229
1970	0.324488
1971	0.374397
1972	0.392473
1973	0.439437
1974	0.537819
1975	0.643884
1976	0.742414
1977	0.813682
1978	0.883572
1979	0.952377

Note : *(a) Base value of 1.0 at 31/12/79. The index is proportional to Average Weekly Earnings for ACT. In years 1973 and earlier, this statistic was not published. It has been taken as 120% of NSW AWE. Inflation of AWE from mid-1979 to end-1979 has been taken as 5%.*

A5. Claim payments in 31/12/79 values (c_{ij}/λ_{i+j}) in the notation of Section 3)

Accident year	Claim payments (31/12/79 values) (a) in development year								
	0	1	2	3	4	5	6	7	8
	$	$	$	$	$	$	$	$	$
1969	193013	1584331	1151882	778980	475203	143352	128612	70845	25077
1970	376473	1541950	1719509	1032570	289305	382508	270087	108354	23133
1971	568891	1579158	1277822	734670	680369	217221	147800	57099	64829
1972	428753	970640	955898	1095771	510072	491853	242995	299845	
1973	458252	989072	1417606	953222	881133	278778	197156		
1974	355229	948807	1292900	748003	547288	274367			
1975	282419	688332	1158793	903450	629983				
1976	267600	1044790	1216437	527644					
1977	560307	940002	1185899						
1978	360171	1011773							
1979	445545								

Note : *(a) Derived from Appendices A3 and A4.*

APPENDIX B

GOODNESS-OF-FIT OF MODELS

B1. Method 1 - Basic chain ladder

Accident year	Claim payments: (1) actually recorded (a); (2) fitted by the model (b); in development year								
	0	1	2	3	4	5	6	7	8
	$	$	$	$	$	$	$	$	$
1969 (1)	57369	514095	418034	305728	208822	77754	82811	52596	20405
(2)	103202	314551	414235	310225	245435	139781	101697	82600	25888
1970 (1)	122161	559595	674860	453750	156919	246291	200516	88166	20440
(2)	149831	456671	601393	450390	356326	202937	147646	119921	37584
1971 (1)	212991	619776	561523	398485	438079	161268	120262	50451	61742
(2)	155882	475114	625681	468579	370716	211132	153608	124764	39102
1972 (1)	168274	426536	518479	705550	378684	400212	214705	285566	
(2)	186783	569298	749712	561467	444205	252986	184059	149496	
1973 (1)	201373	536473	912775	707685	716962	246320	187767		
(2)	222311	677585	892317	668266	528699	301107	219069		
1974 (1)	191049	612854	956866	611637	484568	260301			
(2)	210622	641956	845397	633127	500899	285274			
1975 (1)	181845	511027	942889	798263	599887				
(2)	225638	687725	905671	678266	536611				
1976 (1)	198670	850127	1074810	502516					
(2)	237277	723201	952390	713254					
1977 (1)	455912	830559	1129424						
(2)	299674	913382	1202839						
1978 (1)	318237	963590							
(2)	316664	965163							
1979 (1)	424327								
(2)	424327								

Notes : (a) From Appendix A3.

(b) The fitted value of A_{ij} is obtained as

$$A_{if_i}/M_{jf_i}$$

Fitted values of C_{ij} are then obtained by differencing the A_{ij}.
This procedure ensures that actual and fitted values of A_{if_i} are
equal, i.e. actual and fitted row sums are equal.

B2. Method 2 - Inflation-adjusted chain ladder

Accident year	Claim payments (31/12/79 values): (1) actually recorded (a); (2) fitted by the model (b); in development year							
	0	1	2	3	4	5	6	7
	$	$	$	$	$	$	$	$
1969 (1)	193013	1584331	1151882	778980	475203	143352	70845	25077
(2)	368753	1077985	1193038	795267	526075	266711	118059	32867
1970 (1)	376473	1541950	1719509	1032570	289305	382508	108354	23133
(2)	465380	1360453	1505604	1003654	663925	336599	148994	41479
1971 (1)	568891	1579158	1277822	734670	680369	217221	57099	64829
(2)	433131	1266179	1401318	934105	617917	313274	138669	38604
1972 (1)	428753	970640	955898	1095771	510072	491853	299845	
(2)	407715	1191880	1319089	879292	581658	294891	130532	
1973 (1)	458252	989072	1417606	953222	881133	278778		
(2)	433687	1267804	1403116	935304	618710	313676		
1974 (1)	355229	948807	1292900	748003	547288	274367		
(2)	363413	1062371	1175757	783749	518456	262848		
1975 (1)	282419	688332	1158793	903450	629983			
(2)	340990	996821	1103211	735390	486466			
1976 (1)	267600	1044790	1216437	527644				
(2)	328114	959181	1061554	707622				
1977 (1)	560307	940002	1185899					
(2)	375240	1096946	1214022					
1978 (1)	360171	1011773						
(2)	349690	1022254						
1979 (1)	445545							
(2)	445545							

Notes : (a) *From Appendix A5.*

(b) *As for Appendix B1, but with* C_{ij} *replaced by*

C_{ij}/λ_{i+j} .

B3.   Method 3 - Payments per claim incurred

Accident year	Claim payments (31/12/79 values): (1) actually recorded (a); (2) fitted by the model (b); in development year							
	0	1	2	3	4	5	6	7
	$	$	$	$	$	$	$	$
1969 (1)	193013	1584331	1151882	778980	475203	143352	128612	70845
(2)	315892	927279	1049138	701866	465993	234304	155331	111399
1970 (1)	376473	1541950	1719509	1032570	289305	382508	270087	108354
(2)	388372	1140039	1289858	862906	572913	288064	190971	136959
1971 (1)	568891	1579158	1277822	934670	680369	217221	147800	57099
(2)	408304	1198548	1356056	907192	602316	302848	200772	143988
1972 (1)	428753	970640	955898	1095771	510072	491853	242995	299845
(2)	406492	1193229	1350038	903166	599643	301504	199881	143349
1973 (1)	458252	989072	1417606	953222	881133	278778	197156	
(2)	488636	1434357	1622854	1085678	720819	362432	240273	
1974 (1)	355229	948807	1292900	748003	547288	274367		
(2)	404076	1186137	1342014	897798	596079	299712		
1975 (1)	282419	688332	1158793	903450	629983			
(2)	309852	909549	1029076	688448	457083			
1976 (1)	267600	1044790	1216437	527644				
(2)	327972	962739	1089258	728706				
1977 (1)	560307	940002	1185899					
(2)	375688	1102806	1247732					
1978 (1)	360171	1011773						
(2)	424612	1246419						
1979 (1)	445545							
(2)	448772							

Notes: (a)   *From Appendix A5.*

(b)   *Calculated as :*

$$n_i \hat{v}_j \lambda_{i+j}$$

for accident year i *, development year* j .   *(See paragraph preceding (8.1.4).)*

B.4   Method 4 - Payments per claim finalized (noncumulative)

Accident year	Claim payments (31/12/79 values): (1) actually recorded (a); (2) fitted by the model (b); in development year								
	0	1	2	3	4	5	6	7	8
	$	$	$	$	$	$	$	$	$
1969 (1)	193013	1584331	1151882	778980	475203	143352	128612	70845	20577
(2)	982377	1267266	943152	565656	282828	181398	119077	121770	16148
1970 (1)	376473	1541950	1719509	1032570	289305	382508	270087	108354	23133
(2)	456458	1588197	1574727	599326	259896	371434	459297	219186	72666
1971 (1)	568891	1579158	1277822	734670	680369	217221	147800	57099	64829
(2)	436612	1571739	1625253	525252	756756	423262	170110	48708	24222
1972 (1)	428753	970640	955898	1095771	510072	491853	242995	299845	
(2)	446535	1366014	656838	1245790	1039584	310968	85055	146124	
1973 (1)	458252	989072	1417606	953222	881133	278778	197156		
(2)	515996	946335	2155776	1616160	596232	233226	153099		
1974 (1)	355229	948807	1292900	748003	547288	274367			
(2)	248075	1152060	1818936	868686	535080	267778			
1975 (1)	282419	688332	1158793	903450	629983				
(2)	158768	765297	1061046	760942	542724				
1976 (1)	267600	1044790	1216437	527644					
(2)	158768	938106	833679	592592					
1977 (1)	560307	940002	1185899						
(2)	367151	839358	707364						
1978 (1)	360171	1011773							
(2)	228229	864045							
1979 (1)	445545								
(2)	297690								

Notes :   *(a)   From Appendix A5.*

(b)   Calculated as :

$$n_{ij}\hat{p}_{ij}$$

(see (9.1.9)).

Notes :

(a) From Appendix A5.

(b) By (10.1.9), the fitted values of B_{ik} can be taken as :

$$\hat{\pi}_{jk}\left(\sum_{\alpha=0}^{k} N_{i\alpha}\right)\left[b_{ij}/\sum_{\alpha=0}^{j} N_{i\alpha}\right]$$

for some j.

Now values are fitted by equating actual and fitted row totals. Therefore, with a slight abuse of notation, the above expression can be written (with $j = f_i$) =

$$\hat{\pi}_{f_i k}\left(\sum_{\alpha=0}^{k} N_{i\alpha}\right)\left[b_{if_i}/\sum_{\alpha=0}^{i} N_{i\alpha}\right]$$

The abuse is, of course, the writing of $\hat{\pi}_{f_i k}$ with $f_i > k$. This should be interpreted as $1/\hat{\pi}_{kf_i} = \hat{\pi}_{f_i\infty}/\hat{\pi}_{k\infty}$. Then, the fitted value of B_{ik} is :

$$\left(\sum_{\alpha=0}^{k} N_{i\alpha}\right)\left[b_{if_i}/\sum_{\alpha=0}^{i} N_{i\alpha}\right]\hat{\pi}_{f_i\infty}/\hat{\pi}_{k\infty}$$

The values of B_{ik} are then differenced to produce fitted values of C_{ik}/Λ_{i+k} .

B5. __Method 5 - Payments per claim finalized (cumulative)__

Accident year		Claim payments (31/12/79 values) : (1) actually recorded (a); (2) fitted by the model (b); in development year								
		0	1	2	3	4	5	6	7	8
1969	(1)	193013	1584331	1151882	778980	475203	143352	128612	70845	25077
	(2)	956674	1227249	1027081	542417	309302	212369	153059	114758	8386
1970	(1)	376473	1541950	1719509	1032570	289305	382508	270087	108354	23133
	(2)	494547	1800734	1423947	642665	356456	403790	369919	175695	76135
1971	(1)	568891	1579158	1277822	734670	680369	217221	147800	57099	64829
	(2)	384179	1448711	1569182	419862	747770	413022	185947	122409	36777
1972	(1)	428753	970640	955898	1095771	510072	491853	242995	299845	
	(2)	377760	1204487	626373	1233568	987632	302186	140087	123733	
1973	(1)	458252	989072	1417606	953222	881133	278778	197156		
	(2)	388491	726013	1762476	1407942	504108	217619	168570		
1974	(1)	355229	948807	1292900	748003	547288	274367			
	(2)	195376	956493	1558793	737229	473240	245462			
1975	(1)	282419	688332	1158793	903450	629983				
	(2)	146169	837802	1178026	880285	620695				
1976	(1)	267600	1044790	1216437	527644					
	(2)	178332	1115984	1029296	732859					
1977	(1)	560307	940002	1185899						
	(2)	442574	1200353	1043281						
1978	(1)	360171	1011773							
	(2)	275973	1095971							
1979	(1)	445545								
	(2)	445545								

B6. Method 6 - Reduced Reid

Accident year	Claim payments (31/12/79 values): (1) actually recorded (a); (2) fitted by the model (b); in development year								
	0	1	2	3	4	5	6	7	8
	$	$	$	$	$	$	$	$	$
1969 (1)	193013	1584331	1151882	778980	475203	143352	128612	70845	25077
(2)	990114	1209899	828688	641148	336206	311862	142376	65906	17442
1970 (1)	376473	1541950	1719509	1032570	289305	382508	270087	108354	23133
(2)	636169	1570963	1483836	669366	274810	396020	351212	235419	126095
1971 (1)	568891	1579158	1277822	734670	680369	217221	147800	57099	64829
(2)	539270	1400694	1358154	520056	717056	593439	150957	26461	39891
1972 (1)	428753	970640	955898	1095771	510072	491853	242995	299845	
(2)	529539	1185380	521048	1224294	934280	357515	149570	94203	
1973 (1)	458252	989072	1417606	953222	881133	278778	197156		
(2)	578334	734132	1589883	1434884	514832	244039	79493		
1974 (1)	355229	948807	1292900	748003	547288	274367			
(2)	297145	1022534	1380824	793591	453512	218802			
1975 (1)	282419	688332	1158793	903450	629983				
(2)	250318	901975	1048029	905818	556758				
1976 (1)	267600	1044790	1216437	527644					
(2)	265241	1161925	866955	762393					
1977 (1)	560307	940002	1185899						
(2)	680074	1161382	844803						
1978 (1)	360171	1011773							
(2)	357390	1014514							
1979 (1)	445545								
(2)	445560								

Notes : (a) *From Appendix A5.*

(b) *As in Section 11.1, all working is as for the cumulative payments per claim finalized method (Appendix B5), but with "at end of development year α" translated into "at operational time α". Values of $\hat{\pi}_{k,1}$ for the required operational times k can be found by interpolation of the table of Section 11.2.*

B7.　Method 7 - See-saw method

Accident year	Claim payments (31/12/79 values): (1) actually recorded (a); (2) fitted by the model (b); in development year								
	0	1	2	3	4	5	6	7	8
	$	$	$	$	$	$	$	$	$
1969 (1)	193013	1584331	1151882	778980	475203	143352	128612	70845	25077
(2)	720918	1406174	939232	543396	374070	163065	119700	100150	44300
1970 (1)	376473	1541950	1719509	1032570	289305	382508	270087	108354	23133
(2)	528540	1488609	1250282	722947	351560	432580	319680	141300	167580
1971 (1)	568891	1579158	1277822	734670	680369	217221	147800	57099	64829
(2)	505560	1537741	1354860	707850	710028	506660	166000	40420	62250
1972 (1)	428753	970640	955898	1095771	510072	491853	242995	299845	
(2)	517050	1448516	937560	1270580	639880	412200	77350	101040	
1973 (1)	458252	989072	1417606	953222	881133	278778	197156		
(2)	615680	983480	1601792	1004880	670644	329670	118440		
1974 (1)	355229	948807	1292900	748003	547288	274367			
(2)	313500	909020	1274400	988914	576100	346270			
1975 (1)	282419	688332	1158793	903450	629983				
(2)	206240	703638	1261260	941064	859810				
1976 (1)	267600	1044790	1216437	527644					
(2)	206240	742710	1139490	937200					
1977 (1)	560307	940002	1185899						
(2)	438080	836502	953400						
1978 (1)	360171	1011773							
(2)	296470	904880							
1979 (1)	445545								
(2)	376200								

Notes :　(a)　From Appendix A5.

(b)　For each combination of accident year and development year,
claim payments per claim finalized are fitted by the
regression formula (12.1.4).　The parameters assume the
values displayed in Section 12.1.　The values of the
arguments, operational time and speed of finalization are
given in Appendix C2.　Fitted claim payments are then
obtained as the product of payments per claim finalized
and number of claims finalized (Appendix A2).

APPENDIX C

SEE-SAW METHOD

C1. Payments per claim finalized

Accident year	Payments (31/12/79 values) per claim finalized: (1) actually recorded (a); (2) fitted (b); in development year								
	0	1	2	3	4	5	6	7	8
	$	$	$	$	$	$	$	$	$
1969 (1)	1950	10288	10285	9273	12843	6826	18373	14169	12539
(2)	7282	9131	8386	6469	10110	7765	17100	20030	22150
1970 (1)	8184	8246	9195	11602	8509	8896	10003	12039	2570
(2)	11490	7713	6686	8123	10340	10060	11840	15700	18620
1971 (1)	13338	8268	6621	9419	6872	4433	14780	28550	21610
(2)	11490	8051	7020	9075	7172	10340	16600	20210	20750
1972 (1)	9528	5847	12255	5923	3751	13663	48599	49974	
(2)	11490	8726	12020	6868	4705	11450	15470	16840	
1973 (1)	8813	8601	5538	3972	11297	10325	21906		
(2)	11840	8552	6257	4187	8598	12210	13160		
1974 (1)	14089	6799	5986	5798	7818	8851			
(2)	12540	6493	5900	7666	8230	11170			
1975 (1)	17651	7401	9197	7995	8872				
(2)	12890	7566	10010	8328	12110				
1976 (1)	16725	9165	12287	5996					
(2)	12890	6515	11510	10650					
1977 (1)	15143	9216	14118						
(2)	11840	8201	11350						
1978 (1)	15660	9636							
(2)	12890	8618							
1979 (1)	14852								
(2)	12540								

Notes : (a) *From Appendices A2 and A5.*

(b) *Calculated from regression equation (12.1.4) using parameters of Section 12.2 and observed operational times and speeds of finalization tabulated in Appendix C2.*

C2. Operational times and speeds of finalization

Accident year	(1) Average operational time (a) (c); (2) Speed of finalization (b) (c); during development year										
	0	1	2	3	4	5	6	7	8	9	10
1969 (1)	0.095	0.335	0.585	0.780	0.895	0.950	0.976	0.988	0.994		
(2)	0.19	0.29	0.21	0.16	0.07	0.040	0.013	0.010	0.004		
1970 (1)	0.035	0.220	0.515	0.730	0.825	0.884	0.941	0.970	0.984		
(2)	0.07	0.30	0.29	0.14	0.05	0.067	0.042	0.014	0.014		
1971 (1)	0.035	0.210	0.490	0.710	0.820	0.929	0.975	0.984	0.988	0.993	0.998
(2)	0.07	0.28	0.28	0.12	0.14	0.077	0.015	0.003	0.005	0.005	0.005
1972 (1)	0.035	0.190	0.370	0.565	0.805	0.935	0.964	0.972	0.983	0.993	0.998
(2)	0.07	0.24	0.12	0.27	0.21	0.050	0.007	0.009	0.014	0.005	0.005
1973 (1)	0.030	0.135	0.365	0.670	0.870	0.935	0.955	0.965	0.980	0.993	0.998
(2)	0.06	0.15	0.31	0.30	0.10	0.030	0.010	0.010	0.020	0.005	0.005
1974 (1)	0.020	0.145	0.410	0.665	0.815	0.890	0.915	0.940	0.975	0.993	0.998
(2)	0.04	0.21	0.32	0.19	0.11	0.040	0.010	0.040	0.030	0.005	0.005
1975 (1)	0.015	0.120	0.335	0.570	0.750	0.845	0.885	0.920	0.960	0.985	0.995
(2)	0.03	0.18	0.25	0.22	0.14	0.050	0.030	0.040	0.040	0.010	0.010
1976 (1)	0.015	0.135	0.330	0.500	0.665	0.775	0.840	0.900	0.950	0.985	0.995
(2)	0.03	0.21	0.18	0.16	0.17	0.050	0.080	0.040	0.060	0.010	0.010
1977 (1)	0.030	0.140	0.230	0.420	0.565	0.715	0.815	0.880	0.945	0.985	0.995
(2)	0.06	0.16	0.14	0.12	0.17	0.130	0.070	0.060	0.070	0.010	0.010
1978 (1)	0.015	0.105	0.240	0.350	0.500	0.675	0.785	0.860	0.935	0.980	0.995
(2)	0.03	0.15	0.12	0.10	0.20	0.150	0.070	0.080	0.070	0.020	0.010
1979 (1)	0.020	0.110	0.240	0.350	0.500	0.675	0.785	0.860	0.935	0.980	0.995
(2)	0.04	0.14	0.12	0.10	0.20	0.150	0.070	0.080	0.070	0.020	0.010

Notes (a) *Average of operational times at beginning and end of year.*

(b) *Increase in operational time, i.e. proportion of claims incurred finalized, during year.*

(c) *Entries below the heavy diagonal line are derived from the illustrative predicted operational times given in Section 8.2 (first table) of Taylor [9].*

C3. Projected future numbers of finalizations and payments per claim finalized

Accident year	Projected : (1) Number of claims finalized (a); (2) Payments per claim finalized (b); in development year										
	0	1	2	3	4	5	6	7	8	9	10
1971 (1)										4	3
(2)										21824	22864
1972 (1)									9	4	3
(2)									18438	21824	22864
1973 (1)								8	16	4	4
(2)								15273	16942	21824	22864
1974 (1)							6	26	20	3	3
(2)							12709	11936	14449	21824	22864
1975 (1)						26	16	21	21	5	5
(2)						10373	11509	11643	9876	19433	21514
1976 (1)					92	27	43	22	32	5	5
(2)					8314	10375	9316	11349	11322	19433	21514
1977 (1)				75	106	81	44	37	44	6	6
(2)				11994	9986	8712	9669	10295	10868	19433	21514
1978 (1)			84	70	141	105	49	56	49	14	7
(2)			11927	13514	9468	8764	9671	9242	10721	16941	21514
1979 (1)		104	89	74	149	111	52	59	52	15	8
(2)	7618	11927	13514	9468	8764	9671	9242	10721	16941	21514

Notes : (a) *Derived in accordance with Taylor [9] first table in Section 8.2).   The predicted numbers of future finalizations are needed for illustrative purposes only.   No importance attaches to them in the context of the principle involved in the see-saw method.*

(b) *Derived from (12.1.4) using the parameters displayed in Section 12.2 and future operational times and speeds of finalization from Appendix C2.*

REFERENCES

[1]   Cumpston, J.R., Payments per unit of risk claim models. *General Insurance Bulletin*, 1, 8-12, 1976.

[2]   Finger, R.J.,   Modelling loss reserve developments. *Proceedings of the Casualty Actuarial Society*, 63, 90-105, 1976.

[3]   Johnson, R.L., A generalised statement of the claims runoff process. *Institute of Actuaries of Australia General Insurance Seminar* 2, Broadbeach, November 1980 (to appear).

[4]   Matthews, T.J., The valuation of general insurance claim reserves. *Institute of Actuaries of Australia General Insurance Seminar* 1, Thredbo, December 1978, pp.69-97.

[5]   Reid, D.H., Claim reserves in general insurance. *Journal of the Institute of Actuaries*, 105, 211-96, 1978.

[6]   Sawkins, R.W., Some problems of long-term claims in general insurance. *Transactions of the Institute of Actuaries of Australia and New Zealand*, 336-87, 1975.

[7]   Sawkins, R.W., Analysis of claim run off data - a broad view. *Institute of Actuaries of Australia General Insurance Seminar* 1, Thredbo, December 1978, pp. 30-60.

[8]   Sawkins, R.W., Methods of analysing claim payments in general insurance. *Transactions of the Institute of Actuaries of Australia* (to appear).

[9]   Taylor, G.C., A Reduced Reid method of estimation of outstanding claims. *Institute of Actuaries of Australia General Insurance Seminar* 2, Broadbeach, November 1980 (to appear).

[10]   Taylor, G.C. and Matthews, T.J., Experimentation with the estimation of the provision for outstanding claims in non-life insurance. *Transactions of the Institute of Actuaries of Australia*, 178-254, 1977.

A DIRECT COMBINATORIAL ALGORITHM
FOR CUTTING STOCK PROBLEMS

R.E. JOHNSTON,

Department of Mathematics,
Footscray Institute of Technology, Melbourne.

INTRODUCTION

In many industries solid materials are produced in sizes that are larger than needed by the users of the material.   There are obvious economic advantages in doing this but the practice gives rise to a problem of determining how the material should be cut to obtain the required sizes.   This *cutting stock* or *trim* problem is of considerable importance in deciding the final profitability of the manufacturing venture.

This paper attempts to carefully introduce the problem and its complications, to clearly demonstrate why sophisticated computer-based solutions are necessary and finally present a new direct algorithm which has been implemented in practice.

THE BASIC CUTTING STOCK PROBLEM

We shall concentrate in this paper on one dimensional roll cutting problems with a single stock length L .   In general we address ourselves to problems encountered in the paper industry - particularly in fine or printing grade paper mills.

The basic problem is then as depicted in Figure 1.

Orders are characterised by a length ℓ_i and a number.   For example, in Figure 1 order 1 is for 3 rolls of length ℓ_i .

Figure 2 presents a specific example of a trivial cutting stock problem for which two possible solutions are given.   Note that the solutions are in terms of cutting patterns and their associated utilisation levels.   Solution 2 has exactly the same waste as solution 1, but contains only two *different* cutting patterns.

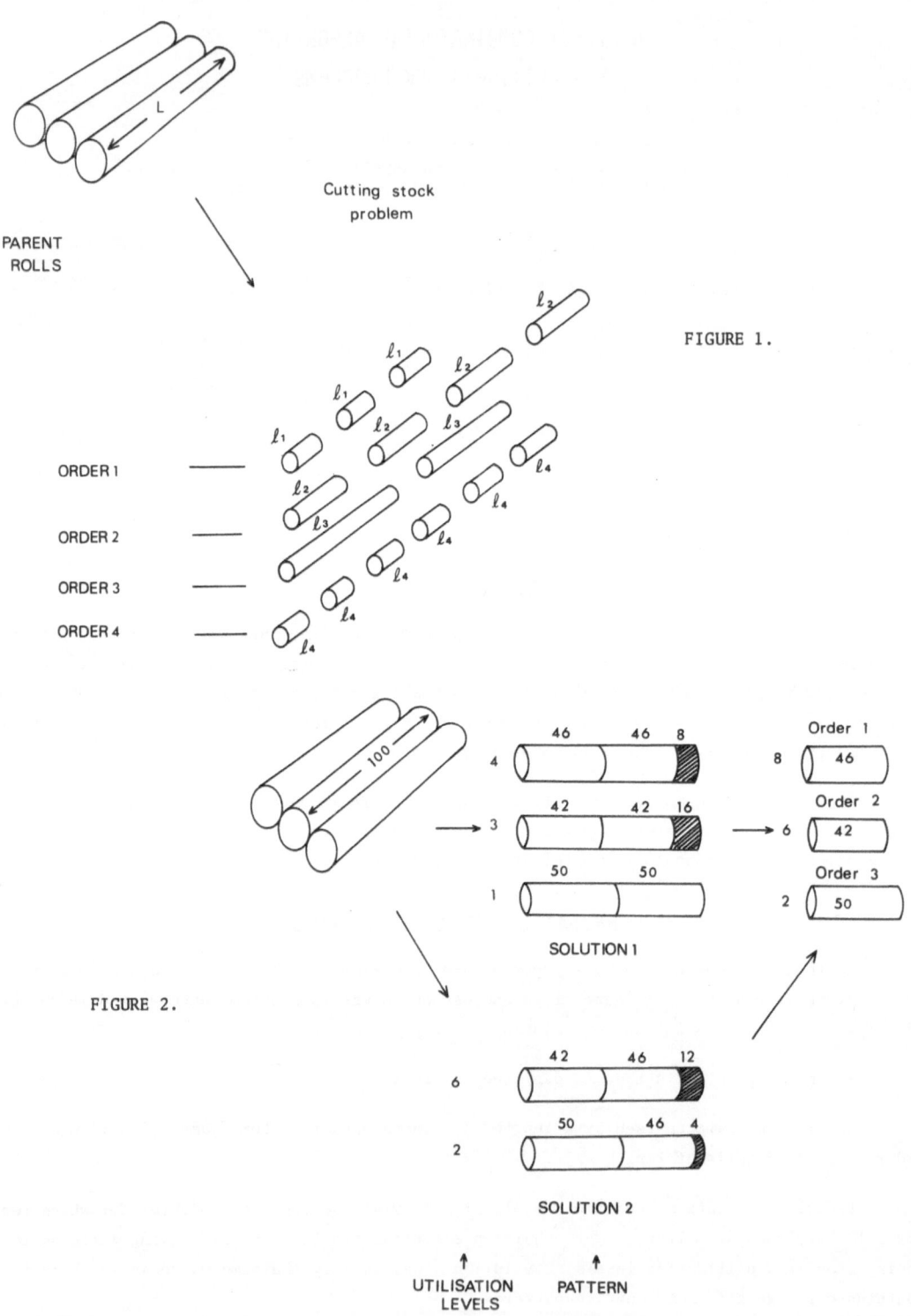

Cutting stock
problem

FIGURE 1.

FIGURE 2.

There are three basic factors of absolute importance in even the simplest form of cutting stock problems.   They are :

(i) Solutions must have integer utilisation levels.   A non-integer solution is generally infeasible as joining or splicing to build up a roll is extremely expensive.

(ii) The solution should have a low waste.

(iii) The solutions must reflect the fact that changing from one pattern to another is both time and labour consuming.   If the rolls are being cut in sequence with another manufacturing process the excessive pattern changes results in bottle-necks at this stage.   Even if the cutting is done off-line, the extra time at pattern changes leads to an increase in time to process a batch of orders.

Before looking at any possible solution procedure, we will discuss the wide range of constraints and conditions which can exist and which make these cutting stock problems extremely difficult combinatorial problems.

VARIATIONS ON THE SIMPLE CUTTING STOCK PROBLEM

Three types of constraints or conditions can be isolated - those that affect the utilisation levels, those that affect the individual patterns and those that specifically affect the sequence of patterns as well as the patterns themselves.

A. The cutting pattern constraints

1. Maximum number of cutting knives.   In some situations the slitting capacity of the cutter may not be sufficient and consideration must be given to reprocessing.   For example, if only 8 rolls can be cut at a time, every pattern that contains more than 8 rolls will be cut into 7 rolls plus 1 extra reprocessing roll.   This roll will be cut further in a reprocessing operation.   An example used by Haesslar [11] is given in Table 1a.

2. Maximum different orders per pattern.   In many instances equipment restrictions or handling problems require that patterns do not contain more than a certain number of different orders.

3. Maximum waste.   Either because of limits on trim handling equipment or because of other requirements, it is very often impractical to allow individual patterns with a waste greater than a specified limit.

4. Orders with position specified.   Very often a particular order will be specified as a "side" order or a "middle" order.   In the first case, it must be cut as the outside roll in a pattern, in the second it must be excluded from the outside positions.

5. Maximum number of rolls of a given order in a pattern.   To spread an order throughout a making, it is often useful to restrict its occurrence in any one pattern.

6. Restricted areas. Because of faults which from time to time occur in the manufacturing process, it is often necessary to arrange patterns so that they will fit around the faulty spot. For example, Figure 3 depicts the situation where a pattern has been designed to "miss" a section near the middle of the parent roll.

7. Succeeding operation constraints. In this case it is essential to generate cutting patterns that simultaneously consider constraints arising in subsequent operations. An example from Haesslar [11] is given in Table 1b to demonstrate this class of problem. Each parent roll is 200 inches wide and must be cut into 3 rolls (for secondary processing), none of which must exceed 72 inches.

8. Optional sizes. Patterns must generally be constructed from ordered sizes but often it is possible to produce other sizes (for stock) in order to obtain less trim.

TABLE 1a
REPROCESSING PROBLEM

Required Sizes and Quantities

25.000	5	13.875	29	10.125	14
24.750	73	12.500	87	10.000	43
18.000	14	12.250	9	8.750	15
17.500	4	12.000	31	8.500	21
15.500	23	10.250	6	7.750	5
15.375	5				

Maximum width 172 inches
Maximum cutting capacity 10 rolls
Reprocessing edge requirement 1.5 inches (i.e. if reprocessing is necessary then an extra 1.5 inches must be allowed).

SOLUTION

PATTERN 1   14 SETS   LOSS   .375/SET
  3   24.750
  1   18.000
  1   15.500
  1   13.875
  4   12.500

PATTERN 2   5 SETS   LOSS   .5/SET
  4   24.750
  1   15.500
  1   15.375
  3   13.875

PATTERN 3   4 SETS   LOSS   2./SET
  1   25.000
  2   24.750
  1   17.500
  1   15.500
  5   12.500

PATTERN 4   7 SETS   LOSS   .5/SET
  1   12.500
  1   12.250
  4   12.000
  2   10.125
  1   10.000
  1   68.500   REPROCESS

```
┌─────────────────────────────────────┐
│ PATTERN FOR 68.5"                    │
│ REPROCESSING ROLLS                   │
│ PATTERN A   7 SETS   LOSS   1.5/SET  │
│   5   10.000                         │
│   2    8.500                         │
└─────────────────────────────────────┘
```

PATTERN 5   3 SETS   LOSS   0
  1   24.750
  1   10.250
  5   8.750
  2   8.500
  1   76.250   REPROCESS

```
┌──────────────────────────────────────────┐
│ PATTERNS FOR 76.250"                      │
│ REPROCESSING ROLLS                        │
│ PATTERN B   1 SET   LOSS   1.5            │
│   3   12.000                              │
│   3    7.750                              │
│ PATTERN C   1 SET   LOSS   1.75           │
│   4   12.500                              │
│   2   12.250                              │
│ PATTERN D   1 SET   LOSS   2.0            │
│   1   25.000                              │
│   3   10.250                              │
│   1   10.000                              │
│   1    8.500                              │
└──────────────────────────────────────────┘
```

TABLE 1b

SEQUENTIAL OPERATIONS PROBLEM

Required Sizes and Quantities

28.625	852	23.750	360
26.000	188	14.250	428
25.000	488	13.625	537

Maximum width primary operation 200 inches
Maximum width secondary operation 72 inches (convertor rolls)
Edge loss at secondary operation .875 inches (included in all convertor rolls).

SOLUTION

PATTERN 1   88 SETS   LOSS   .125/SET
  CONVERTOR ROLL 1   WIDTH 67.500
1 28.625
1 23.750
1 14.250
  CONVERTOR ROLL 2   WIDTH 67.500
1 28.625
1 23.750
1 14.250
  CONVERTOR ROLL 3   WIDTH 64.875
1 26.000
1 23.750
1 14.250

PATTERN 2   4 SETS   LOSS   .125/SET
  CONVERTOR ROLL 1   WIDTH 71.750
2 28.625
1 13.625
  CONVERTOR ROLL 2   WIDTH 64.875
1 26.000
1 23.750
1 14.250
  CONVERTOR ROLL 3   WIDTH 63.250
1 25.000
1 23.750
1 13.625

PATTERN 3   87 SETS   LOSS   .375/SET
  CONVERTOR ROLL 1   WIDTH 71.750
2 28.625
1 13.625
  CONVERTOR ROLL 2   WIDTH 69.750
1 28.625
1 26.000
1 14.250
  CONVERTOR ROLL 3   WIDTH 58.125
2 28.625

PATTERN 4   11 SETS   LOSS   0
  CONVERTOR ROLL 1   WIDTH 68.750
1 28.625
1 25.000
1 14.250
  CONVERTOR ROLL 2   WIDTH 66.125
1 26.000
1 25.000
1 14.250
  CONVERTOR ROLL 3   WIDTH 65.125
2 25.000
1 14.250

PATTERN 5   43 SETS   LOSS   0
  CONVERTOR ROLL 1   WIDTH 68.750
1 28.625
1 25.000
1 14.250
  CONVERTOR ROLL 2   WIDTH 66.750
1 25.000
3 13.625
  CONVERTOR ROLL 3   WIDTH 64.500
2 25.000
1 13.625

PATTERN 6   90 SETS   LOSS   .5/SET
  CONVERTOR ROLL 1   WIDTH 68.125
1 28.625
1 25.000
1 13.625
  CONVERTOR ROLL 2   WIDTH 68.125
1 28.625
1 25.000
1 13.625
  CONVERTOR ROLL 3   WIDTH 63.250
1 25.000
1 23.750
1 13.625

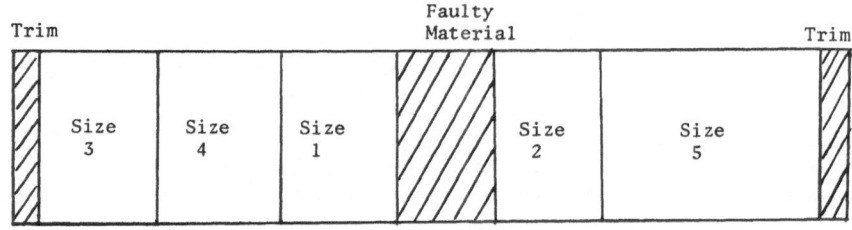

FIGURE 3.

B. Utilisation level constraints

1. Minimum run-length.   Very often a minimum utilisation level is specified in order to avoid bottlenecks occurring due to excessive set-up time over a short period.

2. Utilisation levels in multiples of n .   Occasionally it is desirable to have the utilisation levels, especially the small ones, at a multiple of some number (usually 2 or 3).   This situation arises when a parent roll (diameter) corresponds to an integer number of customer roll (diameter).   For example a parent roll of diameter 70 corresponds to three customer rolls of diameter 40 $(3 \times 40^2 \simeq 70^2)$.

C. Sequencing constraints

Generally cutting stock problems are solved without consideration to the sequencing of the final patterns.   It is simply assumed that one can sequence the patterns in some acceptable way.

In recent years with a trend to automate finishing lines, it has become apparent that one must be able to explicitly control the sequencing as well.

1. Due-date constraints.   If strictly observed due-dates apply to all orders then obviously one must produce a sequence of patterns which as near as possible, satisfies all these due dates.   In a simplistic way, this means that orders with similar due-dates must be grouped together.

2. Work-in-progress limitations.   In many situations there is only a limited area to store partly completed orders.   In this situation the sequence of patterns must be such that the work-in-progress store is not filled.

3. Make-span limitations on individual orders.   These constraints require that a particular order is made - that is started and finished - within a specified time.

This type of constraint obviously avoids the situation where 95% of an order is made early in the making and the last 5% is not made until the last pattern.

SOLVING THE CUTTING STOCK PROBLEM

For at least 100 years, cutting stock problems have been tackled by manual planning methods.   Experienced and obviously talented personnel have battled with the many complex and combinatorially formidable aspects of these problems.

Since 1963 however, when Gilmore and Gomory [7] presented their paper on the cutting stock problem, there have been many attempts to solve these problems, using mathematical algorithms and the power of the computer.   Indeed the paper company is rare that has not had some flirtation with the use of computers to generate cutting schedules.

But why this great desire to change the *modus operandi?*   The fact is that only small percentage improvements in production yields have a significant economic benefit.   Although reported savings from the application of computerised cutting schedules are few in number, those available are extremely encouraging.   (Johnson and Bourke [13], APM [1], Crone [3], Haesslar [9]).

In addition, there are a number of other important benefits.   Apart from giving personnel more time, a computerised system reduces the chance of errors and increases the capability of dealing with last-minute changes.   To management, worrying about their dependence on the manipulative skills of their planning staff, it offers great comfort.

SOLUTION METHODS - A REVIEW

Kantorovitch [16] was the first to consider problems of minimisation of scrap as mathematical programming problems.   He suggested applications in the paper, glass, metal and lumber industries.   Three other early papers (Eisman [4]; Metzger [17]; Paull and Walter [18]) all consider the problem of minimising trim losses when cutting rolls of material to satisfy customer demands.   These demands include only integral numbers of rolls of varying widths. All the possible combinations (patterns) for the setting of the knives are then enumerated. Using the number of times each pattern is used as the decision variables the problem is formulated as a linear program by relaxing the integer constraints on the variables.   This formulation remains as the basis of all the later linear programming trim models.   To obtain an integer solution the results were simply rounded upwards.

The difficulty of this initial approach is that the number of patterns to be considered is enormous.   Gilmore and Gomory ([6],[7]) overcame this problem with a column generation technique enabling one to implicitly consider all the possible patterns while holding only a few at any one time.   In this model a number of different stock lengths ℓ_k (k=1,2,...,n)   are available in unlimited supply to be cut to fill incoming orders.   Each order consists of a demand for N_i pieces of length ℓ_i $(i=1,2,\ldots,m)$. Given that each stock length L_k has an associated cost C_k the problem is to fill the orders at minimum cost. The basic idea of their algorithm is that at every pivot step during the revised simplex procedure the column (pattern) to enter the basis is generated by solving an auxiliary problem. This auxiliary problem is the well known one-dimensional knapsack problem. In their first paper (Gilmore and Gomory [6]) they gave a dynamic programming method for solving this problem while in their second paper (Gilmore and Gomory [7]) they presented an implicit enumeration algorithm which is faster for the paper industry trim problem. Extensions to the model were also presented to allow for

(i)   a maximum number of cutter knives

(ii) limitations on the availability of each stock length (machine capacity)

(iii) tolerance to be allowed on customer demands.

Pierce [19] observed that the linear programming solution without subsequent rounding (for both single and multiple stock situations) is important for solving problems arising naturally when demands are specified lengths or numbers of sheets. Also included are several heuristic procedures for the paper industry trim problem and a review of possible scheduling systems. In a later paper (Pierce [20]) he discusses the solution of the integer constrained problem by first finding a feasible solution and then searching only those remaining solutions which dominate the present one.

Several papers (Hiron [12], Johnston [13]) also discuss the application of the models and methods developed by Gilmore and Gomory. It is shown that the basic multi-machine model can be used to solve several extensions to the basic problem :

(i)   distribution costs can be included where machines are in different locations.

(ii) production for stock can be allowed with trim savings being balanced against increased stock holding costs.

(iii) mixed scheduling of orders with over-lapping quality specifications can be carried out.

Several authors (Filmer [5], Pierce [19]), discuss heuristic methods for specific paper cutting problems. Most interest in heuristic algorithms arose, however, from attempts to solve the non-linear problems arising due to the type of practical constraints mentioned earlier. Most importantly, limits on the number of cutting patterns and on a minimum run length for any pattern cannot be included in the linear programming models. Haesslar [8,9,10, 11] develops heuristic approaches to these problems. Starting with order listed from high demand to low demand, the basic heuristic sequentially generates patterns until a "satisfactory one is found. (This must satisfy certain goals regarding trim loss and number of items in the pattern). This pattern is used to its maximum extent, the order requirements are reduced correspondingly, and the process repeated until all demands are satisfied. In this fashion, patterns are introduced which have a high utilisation and hence the final solution tends to have a relatively small number of cutting patterns. Figure 4 is a schematic of Haesslar's approach. Haesslar [9] reports success in using this algorithm routinely. The algorithm is effective on problems in which order widths are relatively small, compared to the stock widths.

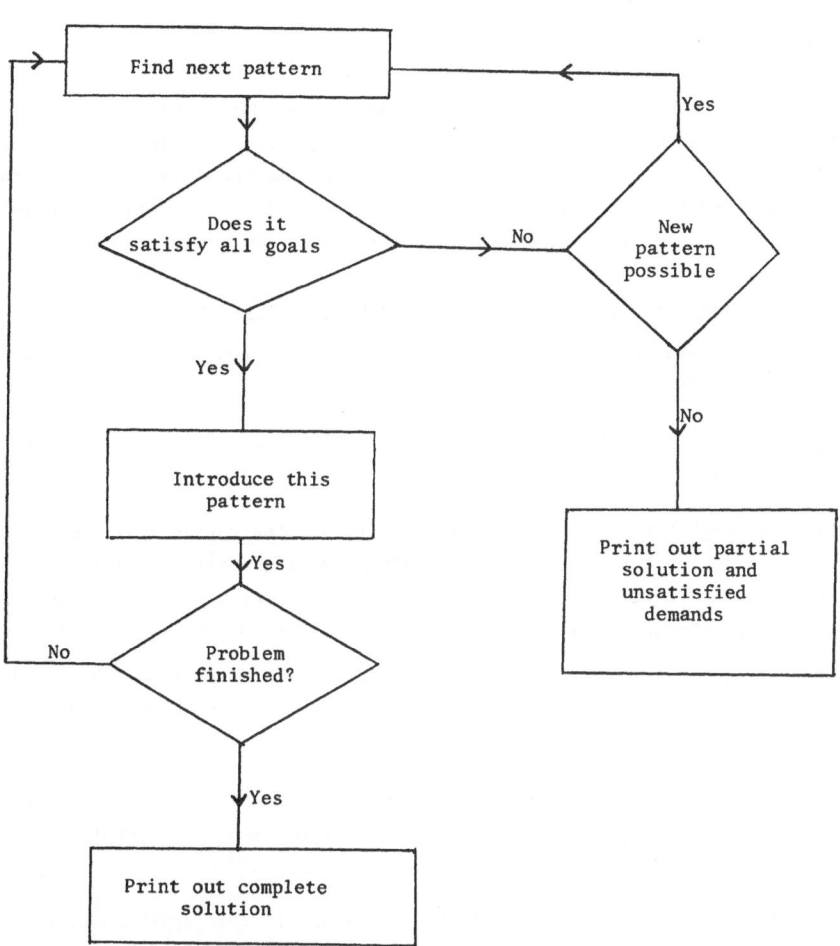

FIGURE 4. Schematic of Haesslars heuristic algorithm.

Further improvements to Haesslars algorithm are presented by Coverdale and Wharton [2] who introduce a check on the remaining problem before introducing patterns. If the remaining problem would necessitate a pattern with run-length too short then the pattern is rejected and a further pattern sought. In addition they improve upon the pattern generation procedure and report better schedules than Haesslar.

SOME NEW DEVELOPMENTS

The introduction of heuristics to overcome some of the non-linear constraints encountered in practice has been a major development in trim programs for the paper industry. This paper presents several new developments which extend the usefulness of these types of approaches.

For the purposes of discussion it is necessary to distinguish two types of trim problems. The class of problems in which the average size of orders is small (averaging less than L/5) in relation to the stock size is known as the *easy* class. (All other problems are called *hard* and the algorithm discussed here does not apply. These problems arise most often in paper or board mills producing packaging grade products).

Extensions to Haesslars Algorithm for the Class of Easy Problems

The key to Haesslars algorithm [9,10] is a pattern generation at each iteration according to the following procedure :

(1)   Sort the remaining orders on the basis of the number of items (reels) still required for each order.

(2)   Find the next pattern by generating patterns in lexicographically decreasing order "with the restriction that the order size with largest quantity still to be scheduled, must be included". Patterns are only accepted if they meet certain goals controlling the quality of the solution.

There exists a simple modification to this pattern generation scheme which appears to be soundly based theoretically. At step (2) of the above procedure carry out the following :

(i)   Start with the maximum size set (utilisation level) possible at this stage. This can be calculated simply from the remaining demands. Denote this utilisation level by MAXNR.

(ii)   Attempt to construct a pattern satisfying all the goals, which can be used MAXNR times. This can be achieved using the following method - Let the set A be the set of indices of those orders which can be used in a pattern at utilisation level MAXNR. Then ℓ_i , $\forall i \in A$ are the widths to be considered. Define f_i as the maximum number of times ℓ_i can be included in a pattern at this utilisation level. Good patterns can be generated by solving the value-independent knapsack problem.

$$\max \sum_{i \in A} a_i \ell_i$$

$$\text{s.t.} \sum_{i \in A} a_i \ell_i \leq L$$

$$\text{and} \quad a_i \leq f_i \quad \forall i \in A$$

$$a_i \quad \text{integer} , \geq 0 .$$

These problems are relatively small as $|A|$ is usually no more than 10 and an implicit enumeration algorithm is used to solve for the first ten patterns. These patterns are checked to find the best satisfying all the goals. If none of the patterns satisfy the goals the search is broadened, but in practice the first ten seem to always contain a satisfactory pattern if one is to be found at all.

(iii) If a satisfactory pattern can be found at step (ii) then include this pattern and proceed to the next iteration unless all orders are satisfied, in which case stop.

(iv) If no satisfactory pattern is found then set MAXNR = MAXNR - 1 and go to step (ii). Continue until a satisfactory pattern is found.

The modification has some similarities to that discussed by Coverdale and Wharton [2] and has given better results (Johnston [14]).

The Algorithm of Haesslar allows all utilisation levels or set sizes to be used. As discussed earlier, there are many practical situations where certain specified set sizes are not acceptable. The most obvious examples are :

(i) where small set sizes are not allowed, i.e. a minimum run length is imposed.

(ii) where small set sizes must be a multiple of some number associated with the size of parent stock. In most practical situations there is also an allowable range in the demanded quantities and this flexibility is utilised, from the outset, by the manual planners to minimise the pattern changes.

Haesslars algorithm has been extended to allow for both these considerations (see Figure 5). The basic idea is that at the pattern acceptance trials for patterns generated by the generating algorithm a further test is added to ensure that the inclusion of the pattern at the associated utilisation level will not leave a remaining demand which cannot be satisfied. In other words, given the remaining number of sets and any known limits on the number of pattern changes, the item cannot be satisfied at any of its acceptable demand levels, without using some pattern at an unacceptable utilisation level. (A simple example is that if set size 1 and 2 are unacceptable, if there are a maximum of two patterns remaining and if there is between 5 and 6 sets remaining then a remaining demand of 8 could not be satisfied). To carry out this test an array is used which is set up on the basis of the unacceptable set sizes and which indicates if a demand can be satisfied or not given a certain number of different patterns and a certain total remaining number of sets. The array is set up for small values of demands and small numbers of remaining patterns as these are the only situations when problems occur. It can be set-up at the start of the algorithm or read in from file for standard situations, and allows rapid checks to be made. This modification is a significant generalisation of the acceptance tests suggested by Coverdale and Wharton [2].

As a result of these further restrictions on the problem the one-pass strategy of Haesslars breaks down on many examples and the algorithm reaches a stage where no acceptable pattern can be found. To cope with this it is necessary to allow back-tracking. That is,

if a satisfactory pattern cannot be found, the last pattern which was introduced is removed
and the calculation recommenced, avoiding, of course, the inclusion of this last pattern.
(In fact, this back-tracking can be controlled by run-time also; if the algorithm is taking
too long to find a complete solution the calculation can be recommenced from a new initial
pattern).

The Appendix gives details of a typical problem and its solution by this algorithm.
Major steps in the solution sequence are outlined showing the essential back-tracking nature
of the process.

While Figure 5 is a schematic of the algorithm, Figure 6 represents a simple interpretation of the algorithm.

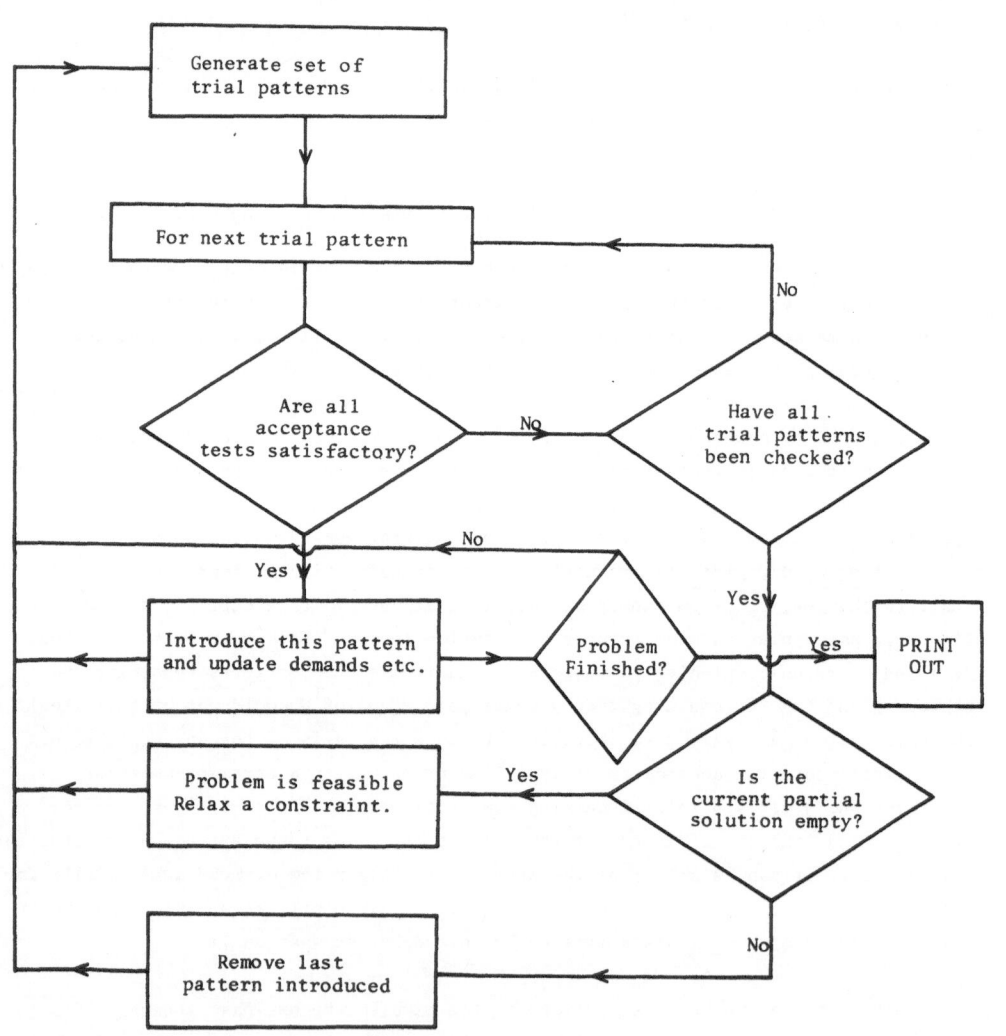

FIGURE 5.

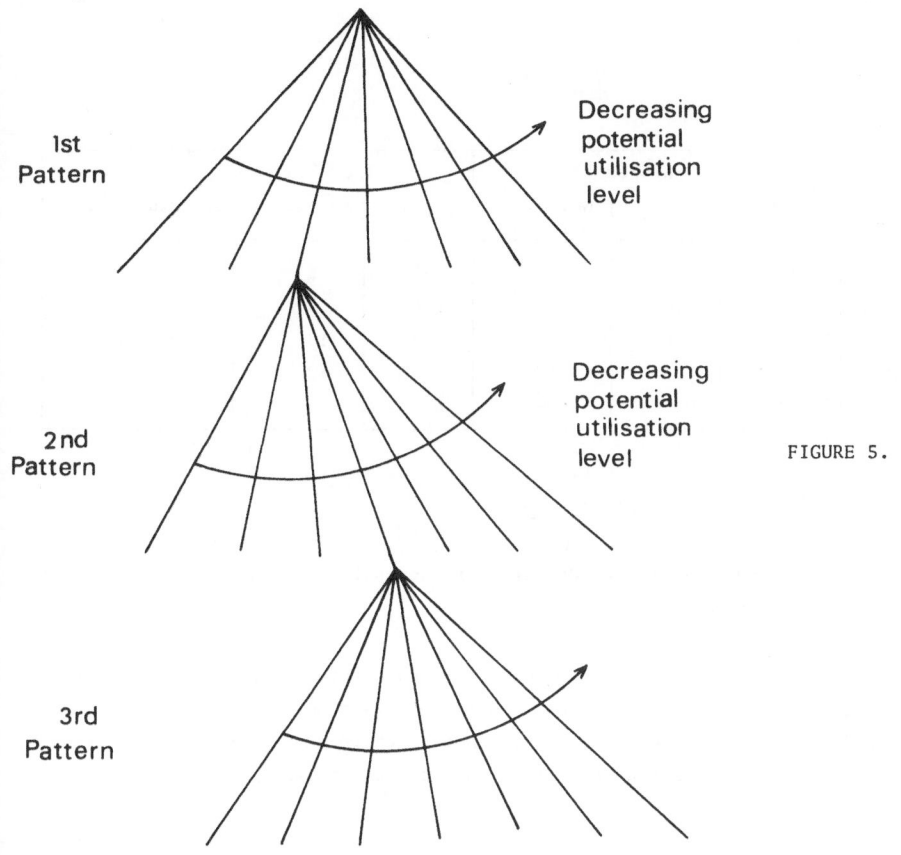

FIGURE 5.

This now puts the cutting stock problem into the format of a branch-and-bound problem. It puts it more firmly where it should most properly be - in the class of *combinatorial optimisation* problems.

Without going into details it is obvious now that all the complex combinatorial constraints can be incorporated in this algorithm.   More details on the bounds used in the calculation are given in Johnston [15].

Computational experience

A new code (FACILE) for constrained cutting stock problems of the easy class has been written.   This code obtains feasible solutions to cardinality constrained, waste constrained cutting stock problems with tolerance.   On problems averaging 20 different items and with a limit of 6 distinct cutting patterns the computational time is of the order of 4 seconds on a CDC 6600 (MNF5).   Several examples are included in the appendix.

A new code for hard problems has also been developed using the additional bounds discussed in Johnston [15].   Computational times are in general similar to those for FACILE.

Table 2 below gives a summary of results obtained on a sample of 5 problems taken from an actual paper mill.   All these problems had constraints on minimum run lengths or other set size limitations.

TABLE 2

Problem	Type	No. of widths	LP result		Tree Search		
			% trim	No. of patterns	% trim	No. of patterns	Computation time sec CDC6600
1	Easy	18	0.0	15	0.2	5	5.6
2	Easy	8	0.0	8	0.07	4	0.2
3	Hard	17	4.8	15	4.5	10	2.5
4	Hard	17	3.35	17	3.43	12	3.7
5	Easy	11	0.1	9	0.3	5	4.1

CONCLUSIONS

The cutting stock problem and its treatment over the past 20 years, highlights several of the themes of this conference.

Firstly, it is a problem which crops up daily in a manufacturing process - each day with a new combinatorial quirk.   What we must offer is not one answer but a method, capable of coping with all these variations.   Secondly, it is an area where many people have recognised a potential for savings to be made but no solution was available to them.   A "problem without a method".

Thirdly, despite the fact that some relatively sophisticated mathematics is necessary, it is basically a simple search which is now providing us with a successful algorithm.

Finally, it is a wonderful example of a problem in which the wrong question was answered by some very nice mathematics.   Perhaps blinded by this breakthrough, it has taken us 20 years to come up with an optimising technique which solves, or at least partly solves the real problem.

REFERENCES

[1]   APM Annual Report, 1972. APM Ltd, Melbourne.

[2]   Coverdale, I. & Wharton, F. An improved heuristic procedure for a non-linear cutting
          stock problem. *Man. Sci.* 23, 1976, 78-86.

[3]   Crone, J.M. *NZFP Paper Trim System* 1972. N.Z. Forest Products Ltd., Auckland, N.Z.

[4]   Eismann, K. The trim problem. *Man. Sci.* 3, 1957, 279-284.

[5]   Filmer, P.J. Duplex cutter deckle filling. *Appita,* 1970, 24, 189-196.

[6]   Gilmore, P.C. & Gomory, R.E. A linear programming approach to the cutting stock
        problem. *Oper. Res. (USA)*, 9, 1961, 849-859.

[7]   Gilmore, P.C. & Gomory, R.E. A linear programming approach to the cutting stock
        problem - part II. *Oper. Res. (USA)*, 11, 1963, 77-82.

[8]   Haesslar, R.W. An application of heuristic programming to a non-linear cutting stock
        problem occurring in the paper industry. *PhD Dissertation*, 1968, Univ. of Mich.

[9]   Haesslar, R.W. A heuristic programming solution to a non-linear cutting stock problem.
        Man. Sci. 17, 1971, B793-B802.

[10]   Haesslar, R.W. Controlling cutting pattern changes in one-dim ensional trim problems.
        Oper. Res. (USA), 23, 1975, 483-493.

[11]   Haesslar, R.W. A survey of one dimensional single stock size cutting stock problems
        and solution procedures, ORSA/Tims meeting, Chicago, 1975.

[12]   Hiron, A.M. Experiences with trim problem. *Quart. Bull. of Brit. Pqp. and Board
        Industry Res. Ass., June, 1966.*

[13]   Johnston, R.E. & Bourke, S.B. The development of computer programs for reels deckle
        filling. *Appita* , 26, 1971, 444-448.

[14]   Johnston, R.E. Extensions to Haesslars heuristic for the trim problem. *Centre
        Technique du Papier,* 1979, Grenoble, France, C.R. No.1366.

[15]   Johnston, R.E. Bounds for the one-dimensional cutting stock problem. Submitted for
        publication.

[16]   Kantorovitch, L.V. Mathematical methods of organising and planning production.
        Leningrad State University and reprinted, *Man. Sci.* 6, 1939, 366-422.

[17]   Metzger, R.W. Stock Slitting. Chap. 8 of *Elementary Mathematical Programming*, 1958,
        Wiley, N.Y.

[18]   Paull, A.E. & Walter, J.R. The trim problem : an application of linear programming
        to the manufacture of Newsprint. *Econometrica,* 23, 1954, 336.

[19]   Pierce, J.F. *Some Large scale production scheduling problems in the paper industry,*
        1964, Prentice-Hall, N.J.

[20]   Pierce, J.F. *On the solution of integer cutting stock problems by combinatorial
        programming. Part I.* 1967, IBM Cambridge Sci. Centre Report.

APPENDIX

Summary of Production

* 5 distinct patterns
* 0.2% waste

Size and quantity produced			
53.0	6	35.0	8
50.0	25	33.5	14
45.0	47	33.0	26
43.5	20	30.0	78
42.0	77	28.0	20
40.0	8	27.0	26
39.0	87	26.0	14
38.0	66	25.0	14
37.0	9	24.0	14

Iteration 1 – Pattern No.1
3 × 39/2 × 42/33/3 × 30
Used 26 times

Iteration 2 – Pattern No.2
50/2 × 45/43.5/42/33.5/28/26/25/24
Used 14 times

Iteration 3 – Pattern No.3
50/2 × 45/2 × 42/2 × 38/35/27
Used 8 times

Iteration 4 – No suitable pattern
Remove Pattern No.3

Iteration 5 – Pattern No.3a
42/40/5 × 38/35/2 × 27
Used 8 times

Iteration 6 – No suitable pattern
Remove Pattern No.3a

Iteration 7 – Pattern No.3b
50/2 × 45/42/40/2 × 38/35/27
Used 8 times

Iteration 8 – Pattern No.4
53/39/4 × 38/27/3 × 27
Used 6 times

Iteration 9 – Pattern No.5
50/45/2 × 43.5/42/39/37/2 × 28
Used 3 times

Example

Data

Required sizes and range of quantities					
53.0	6- 7	:	35.0	7- 8
50.0	25-26	:	33.5	13-14
45.0	45-47	:	33.0	25-26
43.5	19-20	:	30.0	76-81
42.0	77-82	:	28.0	20-21
40.0	8- 9	:	27.0	25-26
39.0	85-90	:	26.0	13-14
38.0	66-69	:	25.0	13-15
37.0	9-10	:	24.0	14-15

- Maximum width 362, maximum of 15 across

- Maximum of 5 patterns

- Set sizes 1, 2, 4, 5 and 7 unacceptable

- Maximum allowable trim, 0.25%

ACCEPTANCE SAMPLING WITH AN EXAMPLE
FROM THE PEANUT INDUSTRY

G.H. Brown,
Division of Mathematics and Statistics,
CSIRO, Sydney.

1. INTRODUCTION

Peanuts, amongst other crops, are subject to attack by fungi of various species; some of which produce toxic substances (aflatoxins); see Schuller [3] for details. Both the Peanut Marketing Board and the manufacturers of peanut products are concerned with monitoring aflatoxin levels. There is scope for the application of quality control methods and, in particular, acceptance sampling plans to assist in this objective. More specifically, following consultation with the Queensland Peanut Board, several problem areas emerged :

(a) What is a reasonable theoretical basis for the distribution of aflatoxin levels ?

(b) To quantify the variability in sampling and to compare results with data from the U.S.A.;

(c) To use the above results in the formulation of sampling plans.

Whilst considerable work has been done in the U.S.A. on such questions there is the need to verify the models and estimates of variability for the Australian Industry. The aim of this paper is to outline the basis of acceptance sampling in the context of the application, although several information gathering experiments have also been recommended.

Sections 2,3 and 4 are concerned with general definitions and objectives (Owen [2], Wetherill and Chiu [4],[5]), whilst section 5 gives details of the peanut application (Schuller et al [3], Whitaker et al [6]). In section 6 a list of difficulties that frequently occur in the application of such plans is given whilst section 7 gives the references for further reading.

2. WHAT IS ACCEPTANCE SAMPLING?

Acceptance sampling refers to the procedure of accepting or rejecting a lot (or batch) of goods on the basis of inspecting (sampling) a portion of the lot.

Acceptance sampling may be seen as the consumer oriented part of quality control. Quality control methods are used by manufacturers to monitor their production; trends and departures from "normal" running conditions are, hopefully, detected and rectified. With acceptance sampling the consumer has given quality specifications and applies the sampling scheme in order to *reduce* the chance that these specifications will not be met.

3. ERRORS RESULTING FROM THE DECISION RULE

The decision rule provides the information of when to accept or reject a lot on the basis of the sample information.   When applied to a lot it gives rise to two possible errors :

		Sample quality	
		Good	Poor
Lot	Good	✓	"Producer" error
Quality	Poor	"Consumer" error	✓

The rate of producer's errors when a given decision rule is applied to a series of lots of acceptable quality is called the producer's risk i.e. the proportion of times that good quality lots will be rejected because the sample results indicated they were poor quality lots. A similar definition holds for the consumer's risk and certain decision rules are based on the pre-specification of these risks.   More generally, the economic consequences of wrong decision need to be known, in which case an attempt is made to balance the consumer's and producer's risks with costs, in order to produce economically viable schemes.

4. POSSIBLE BENEFITS OF ACCEPTANCE SAMPLING

Possible benefits of acceptance sampling are :

(a) Improving the average quality of output by the rejection of poorer quality input lots.

(b) As an incentive for producers to improve input quality e.g. by the introduction of quality control measures.

5. AN EXAMPLE OF AN ACCEPTANCE SAMPLING SCHEME

This example is a highly simplified version of reality.   Hopefully, it illustrates the basic method and, in particular, highlights the nature of the statistical information that is required for the full utilisation of acceptance sampling methods.

A receiving depot obtains peanuts from many farms and each truck-load is tested, using a quick but not accurate test of the aflatoxin level, for the purpose of separating out loads that are potentially highly contaminated. The farm origin of the nuts is then lost as the peanuts are fed into a large storage bin from which they are processed (by shelling and discarding discoloured and other suspect kernels) into bags for dispatch. Just before bagging, a continuous sample is retained and identified with each 10 *tonne* portion of the production which, for testing purposes, becomes the lot. One possible decision rule is the following 3-stage scheme :

Stage/Sample	Decision rule based on mean aflatoxin level ($\bar{x}$)	
	Accept	Reject
1	x ≤ 7 ppb	x > 45
2	x ≤ 13	x > 23
3	x ≤ 15	otherwise

Each sample represents an 8 Kg sample of nuts which are finely ground and then sub-sampled for chemical testing. If the lot quality is high (i.e. aflatoxin level is low) then the scheme will usually only require one test per lot. The second and third stages of the decision rule are designed to give extra protection for intermediate qualities.

The negative binomial distribution has been recommended for the distribution of aflatoxin levels (expressed as parts per billion, ppb) and this, together with the above decision rule, enables the O.C. curve to be calculated. This curve shows the relationship between the probability of accepting a lot (P_A) under the decision rule as a function of the true mean (M) quality of the lot. E.g. - for the decision rule just given :

	$M \left(\begin{array}{c} \text{Mean} \\ \text{quality} \end{array} \right)$	$P_A \left(\begin{array}{c} \text{Probability of accepting} \\ \text{the lot} \end{array} \right)$
	0	1
Good quality	5	.96
	10	.89
	15	.79
poor quality	25	.67
	50	.60

To carry the example further, we examine the performance of the above scheme when the incoming lot quality has the following distribution :

M	0	5	10	15		25	50
Percentage of incoming lots	32	36	17	8		3	4
			93				7

With this information we may calculate the percentage of good lots accepted by the scheme,

$$P_{GA} = 32 \times 1.0 + 36 \times .96 + 17 \times .89 + 8 \times .79 = 88 ,$$

and similarly for other percentages given below :

	Percentage of production	
True quality	A: Accepted	R: Rejected
Good (p_G = 93%)	P_{GA} = 88%	P_{GR} = 5%
Poor (p_p = 7%)	P_{pA} = 2.4%	P_{pR} = 4.6%

i.e. the proportion of correct decisions is $P_{GA} + P_{BR}$ = 92.6%.

For evaluating and comparing various sampling plans the cost information is used in conjunction with the outcome probabilities P_{GA} , P_{BA} , etc. to evaluate the net average cost effect of applying a particular plan.

Cost information relates to :

a. The overhead cost of running the sampling-testing unit.

b. The reduced value of kernels ground up for testing purposes.

c. The cost of recycling lots rejected by the scheme.

d. The cost of recycling lots rejected by the *consumer* (i.e. in this case the manufacturer of nut products).

e. The community costs associated with health risks by a poor quality lot being undetected.

6. DIFFICULTIES IN APPLICATION AND POINTS FOR DISCUSSION

Aspects of acceptance sampling which require further consideration are :

a. What constitutes a "lot" when an acceptance scheme is applied to continuous production?

b. The statistical validity (i.e. the accuracy of the risk assessment) of a sampling scheme usually relies upon a random selection of items to make up the sample. When goods are stored in containers it may be impractical to use random sampling.

c. The dichotomy into good or poor items is simplistic. Often quality measures are on a continuous scale and several characteristics may be involved.

d. Evaluating the risks of a given decision rule depends upon the statistical models used. Sometimes such assumptions will require experimental validation.

e. The cost effectiveness of an acceptance scheme may be extremely difficult to evaluate. Usually the cost of accepting a bad lot is difficult to quantify, especially if health matters are involved. Another factor that arises in cost effectiveness studies is the distribution of incoming quality (the "process" curve).

Hamaker [1] gives an informative example : A number of lots (527) were subject to acceptance sampling inspection based upon MIL.ST. 105A which for the given specifications, prescribed a sample size of n = 86 . The same lots were then sampled with n = 10 using the alternative decision rule : Accept the lot if all ten items meet specification and reject otherwise. The results were :

		Alternative scheme (n = 10)	
		Accept	Reject
MIL.ST. 105A	Accept	449	7
(n = 86)	Reject	7	64

The outstanding success of the alternative scheme in this example results from the distribution of incoming lot quality which was either very good or very poor with few lots of intermediate quality. This type of process curve, i.e. where quality is very very good or horrid, is not likely to be appropriate in many situations. Hence, the above example should not be construed as a recommendation for general application of the alternative sampling plan.

7. REFERENCES

[1]   Hamaker H.C. (1958) Some Basic Principles of Sampling Inspection by Attributes. *Appl. Statist.* 7, 149.

[2]   Owen, D.B. (1969) Summary of Recent Work on Variables Acceptance. Sampling with Emphasis on Non-normality. *Technometrics* 11, 631.

[3]   Schuller, P.L., Horwitz, W. and Stoloff, L. (1976) A review of aflatoxin methodology. *Journal of the AOAC* 59, 1315.

[4]   Wetherill, G.B. and Chiu, W.K. (1974) A Simplified Attribute Sampling Scheme. *Appl. Statist.* 23, 143.

[5]   Wetherill, G.B. and Chiu, W.K. (1975) A review of Acceptance Sampling Schemes with Emphasis on the Economic Aspect. *Inst. Statist. Rev.* 43, 191.

[6]   Whitaker, T.B., Dickens, J.W., Monroe, R.J. and Wiser, E.H. (1972) Comparison of the Observed Distribution of Aflatoxin in Shelled Peanuts to the Negative Binomial Distribution. *J. Am. Oil Chem. Soc.* 49, 590.

ON THE GRINDING OF CONTACT LENSES –
A MATHEMATICAL ABERRATION

A.W. DAVIS,

Division of Mathematics and Statistics,
CSIRO, Adelaide.

ABSTRACT

A commercial firm encountered problems in the grinding and polishing of conicoid contact lenses. The base curves of the lenses are plunge cut into cylindrical blanks on a lathe, and some simple coordinate geometry relating the parameters of the incised curve to the lathe settings showed that an elementary mathematical error had occurred in the specifications for the lathe.

1. DESCRIPTION OF THE PROBLEM

This simple problem was referred to us by a commercial firm engaged in the manufacture of contact lenses. The plastic material from which the lenses are made is initially available in the form of cylindrical buttons, or blanks, each having diameter 13.20 mm and thickness 5.25 mm. In the first stage of the manufacturing process, the blanks are mounted on a lathe fitted with a cutting tool, which plunge cuts a concave surface into the blank (Figure 1(a)). In accordance with the original specifications for the lathe, the cutting tool is a circular steel cylinder sectioned at an angle of 55° to its axis, so that the actual cutting edge is the circumference of an ellipse. The tool is set at an angle to the axis of the lathe, and the surface ground out is in fact conicoidal, being an ellipsoid, paraboloid or hyperboloid of revolution depending upon the choice of angle. These surfaces form the *base curves* of the contact lenses (Figure 1(b)), designed to fit against the cornea of the eye. In the particular process considered, these base curves constitute a standard series of ellipsoids, each having eccentricity 0.7, with radii of curvature at the apices lying in the range 6.08 to 7.75 mm, at an average step size of 0.08 mm. According to the lathe specifications, these requirements should be fulfilled by using cutting tools of appropriate diameters, set at an angle 68.5° to the vertical.

GRINDING OF CONTACT LENSES

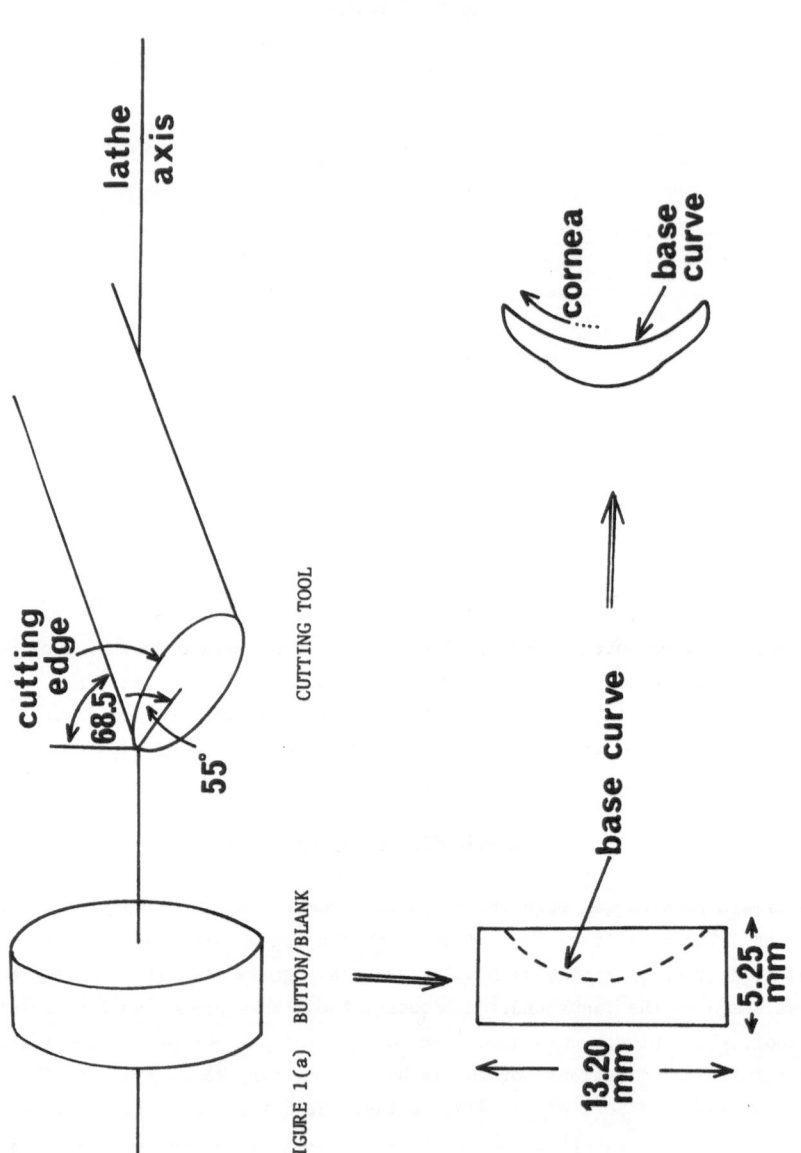

lathe axis

cutting edge

68.5°

55°

CUTTING TOOL

BUTTON/BLANK

FIGURE 1(a)   BUTTON/BLANK

base curve

5.25 mm

13.20 mm

'FIGURE 1(b)   CONCAVE BLANK

cornea

base curve

FIGURE 1(c)   CONTACT LENS

The second stage involves the polishing of the base curves, which is carried out in three steps : (i) the apical area, (ii) the intermediate area between apex and periphery, and (iii) the peripheral area. This is done using padded bronze polishing heads which have been designed to polish the particular range of base curves. Eventually, of course, the polished concave blanks are to be made into contact lenses by grinding anterior surfaces according to individual prescriptions. The particular problem encountered, however, arose at the second stage of the manufacturing process. In theory, the three steps of the polishing should merge smoothly into each other, but in practice it was found that "ripples" were occurring along the regions of overlap. As a result, a significant part of the company's time was being wasted in ad hoc polishing and grinding of the surfaces until they conformed with quality control standards. It was felt that something might be amiss with the lathe settings at the first stage. The patent for the lathe was held by an overseas company, which was approached in the matter, but refused to part with any information concerning the theory relating the lathe settings to the parameters of the incised curve. At this point, the manufacturing firm decided to seek assistance from C.S.I.R.O.

2. THE MATHEMATICAL PROBLEM

The problem of the incised curve simply amounts to deriving the surface of revolution of an ellipse which is inclined at a constant angle to the axis of rotation, with one vertex fixed to this axis. As a first step, we note that the equation for a conic may be conveniently expressed in terms of the parameters of interest by taking a vertex on the major axis as origin. The case of an ellipse is indicated in Figure 2(a). It is easily checked that, if a , b are the semi-major and minor axes respectively, then

$$y^2 = 2rx - (1-e^2)x^2 , \tag{1}$$

where $r = b^2/a$ is the radius of curvature at 0 , and $e = (1-b^2/a^2)^{\frac{1}{2}}$ is the eccentricity. Equation (1) remains true for the parabola $(e = 1)$ and hyperbola $(e > 1)$. For an elliptical cutting edge (Figure 2(b)), inclined at an angle α to the axis of a tool of radius ρ ,

$$a = \rho/\sin \alpha , \quad b = \rho$$

whence

$$r = \rho \sin \alpha , \quad e = \cos \alpha . \tag{2}$$

Referring to Figure 3, let OX coincide with the lathe axis, and let OY lie in the plane containing OX and the major axis of the cutting edge. Suppose that $P(X,Y)$ is a point on the incised curve in the XOY plane, cut by the point Q on the cutting edge; thus, P , Q , G and H lie in a plane perpendicular to XOY . Then

$$Y^2 = GQ^2 = GH^2 + HQ^2 . \tag{3}$$

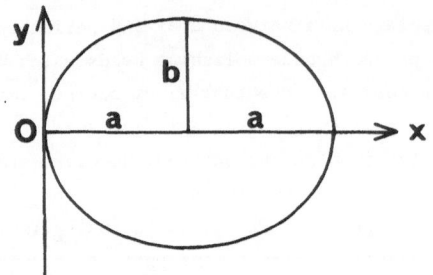

FIGURE 2(a) : Ellipse with origin
at vertix.

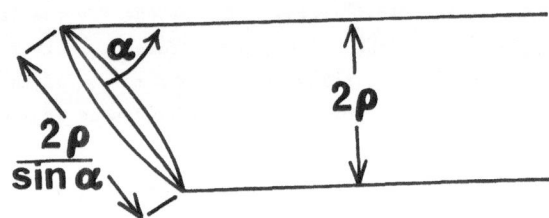

FIGURE 2(b) : Cutting edge.

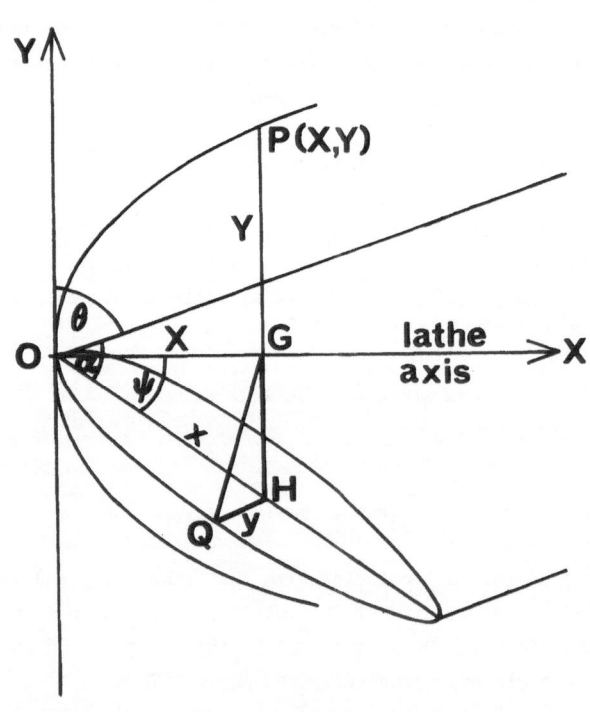

FIGURE 3 : Geometry of the base curve.

If θ is the inclination of the cutting tool to OY, let

$$\psi = \theta + \alpha - 90° \tag{4}$$

denote the angle between OC and the cutting face. Also let $OH = x$, $HQ = y$. Then from (1) and (3)

$$Y^2 = X^2\tan^2\psi + y^2$$

$$= X^2\tan^2\psi + \{2rx - (1-e^2)x^2\} .$$

But $x = X \sec \psi$, and so

$$Y^2 = X^2\tan^2\psi + \{2rX \sec \psi - (1-e^2)X^2\sec^2\psi\}$$

$$= 2r \sec \psi.X - (1-e^2 \sec^2\psi)X^2 . \tag{5}$$

Hence from (1), the incised curve is a conicoid, with eccentricity

$$E = e \sec \psi = \cos \alpha/\cos \psi \tag{6}$$

and radius of curvature at the apex 0

$$R = r \sec \psi = \rho \sin \alpha/\cos \psi , \tag{7}$$

in virtue of (2). We note from (6) and (7) that the eccentricity of the incised conicoid is completely determined by the angles α and θ; with these angles fixed, the apical radius of curvature is seen to be proportional to the radius of the cutting tool.

3. SOURCE OF THE ERROR

The anomaly observed in the grinding and polishing of the lenses is now readily explained using (6). For $\alpha = 55°$, an eccentricity $E = 0.7$ is achieved by taking

$$\cos \psi = \cos 55°/0.7 = 0.81939 .$$

Reference to 5-figure tables shows that

$$\cos 35° = 0.81915 ,$$

with a difference 0.00024 corresponding to an angle of 1.5 minutes which must be subtracted. Thus, from (4),

$$\psi = \theta + 55° - 90° = 35° - (1.5 \text{ min})$$

i.e.
$$\theta = 70° - (1.5 \text{ min}) \ .$$

As indicated in Section 1, the lathe specifications required the cutting tool to be set at an angle $\theta = 68.5°$. Plainly, 5-figure tables had been used in the calculations, and the error arose because 1.5 *degrees* were subtracted instead of 1.5 minutes!

THE FABRICATION OF A THREE DIMENSIONAL
FEEDER: A PROBLEM IN SHEET-METAL BENDING

D. CULPIN,
Division of Mathematics and Statistics,
CSIRO, Sydney,

AND

R. COWAN,
SIROMATH Pty Ltd, Sydney.

ABSTRACT

Here is a problem which resulted from enquiries by an engineering firm and concerns the bending of a piece of sheet metal to a specified shape. The shape is that required of part of a packaging machine. This article may be of interest (a) as a simple illustration of the use of mathematics in abstracting certain features from the real world, (b) for the mathematics, which is an exercise in differential geometry. The article also illustrates that at times a mathematical approach is essential.

1. INTRODUCTION - THE PROBLEM

Figures 1 and 2 show part of a packaging machine which seals loose material, such as powders, into plastic bags. We shall call this part a 'feeder'. Both the material to be packaged and plastic sheeting for the bags are fed into the machine through the feeder. The feeder consists of a 'collar' and a cylindrical part (see Figures 1 and 2). The plastic sheet passes upwards over the collar of the feeder, across the top edge and down into the cylinder, where it is made into bags and filled.

Different sizes of bag require different feeders; these are made to order in Germany. As they are expensive and it is time-consuming to ship them to Australia, an Australian company which imports packaging machines undertook to make their own feeders. They considered that the quickest way would be by bending a piece of sheet metal into a cylinder, at the same time turning the collar over along a curve marked by a groove on the back. Figure 3 shows the piece of sheet metal that could be bent to produce the feeder shown in Figures 1 and 2.

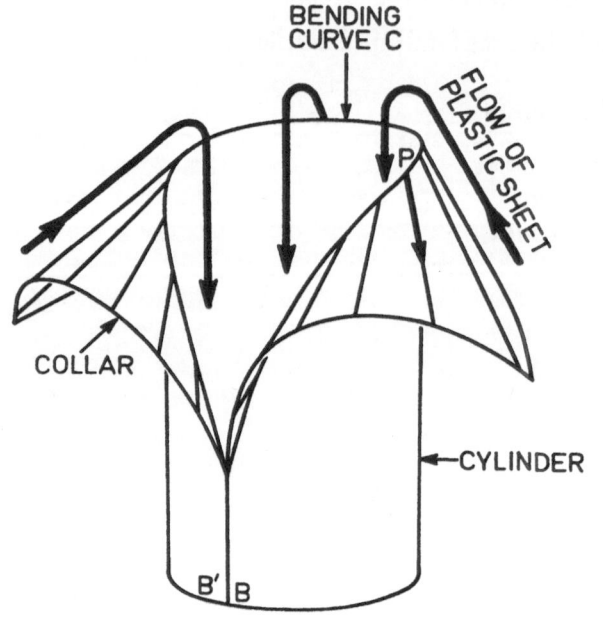

FIGURE 1.

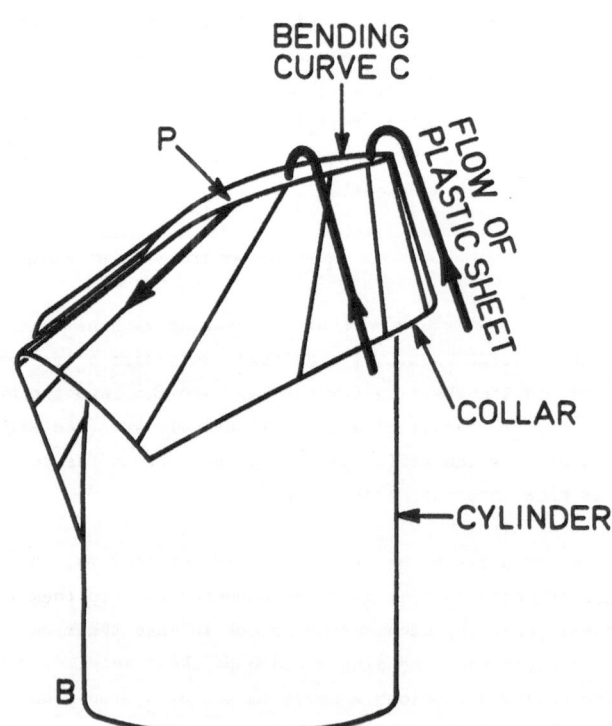

FIGURE 2.

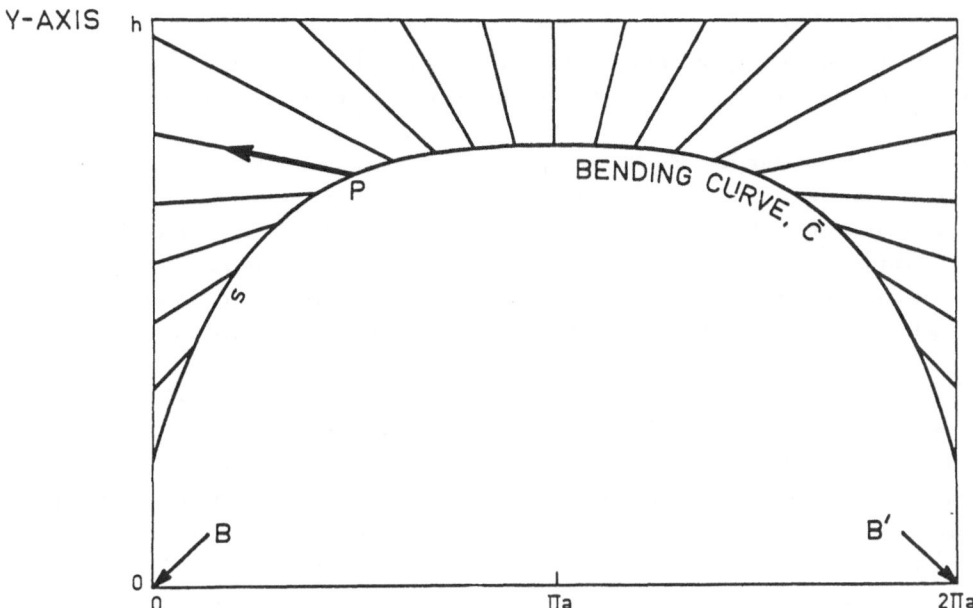

FIGURE 3.

A problem that faced the company was finding the curve along which the collar should be bent in order to produce a feeder of the required shape. Let us call such a curve a 'bending curve'. An obvious requirement of a bending curve is that it be smooth. The main requirements of the shape of the feeder are that the angle between the collar and the cylinder at the highest point should be about 15° and the vertical distance from the lowest to the highest point of the opening should be about 2½ times the radius of the cylinder. The company tried bending curves which were assembled from segments of circles, but their endeavours were not entirely successful, as distortions and tearing occurred along the bending curve when the collar was turned (or rather forced) down to the required angle. They then sought the help of the CSIRO.

2. THE SOLUTION

On considering the situation, questions that will come to mind are : Will any sufficiently smooth curve suffice as a bending curve (provided, of course, that it produces a feeder of the required shape)? Does the bending curve determine the angle between the collar and the cylinder? To answer these questions we must seek mathematical expression for the physical constraints that are operating. These constraints are that there should be no stretching or tearing of the material of the collar; that is, there should be no deformation of this surface in any plane tangential to it. Mathematical expression of this is found in the idea of isometric correspondence. A one-to-one correspondence between two surfaces or curves is isometric when the distances between any corresponding pairs of points in the surfaces or curves are identical.   By 'distance' is meant distance within the surfaces or curves; in the case of curves, distance is arc length.   Because areas and angles can be defined in terms of distances, they also are preserved in an isometric correspondence.

Let S be the collar and $\bar{S}$ its plane form before bending.   Let $\bar{C}$ be the plane bending curve which lies .in $\bar{S}$ and C the bending curve when it lies in S .   The mathematical requirements of the situation can now be stated as follows.   There must exist a surface S which contains C and corresponds isometrically to $\bar{S}$ in such a way that the correspondence includes that of C to $\bar{C}$.   Of course, we know that there is one such surface, namely the cylinder; in order to solve the problem it is necessary to find a second surface.

The question of the existence of the surface S is answered within the subject of differential geometry and involves the notion of curvature for curves in space.   For a space curve, such as $\bar{C}$ in Figure 3, which lies in a plane, the notion of curvature at a given point P is simple; the *radius of curvature* at P is the radius of the circle which best approximates the curve at P .   The *curvature* $\bar{\kappa}$ (of our $\bar{C}$) is the reciprocal of this radius.   For a curve such as C not lying within one plane we can also define a curvature κ (see Weatherburn, [2]).   Basically, a point on the space curve can be represented in three dimensional coordinates $\underline{r}(s)$ indexed by the distance s along the curve.   The derivative of $\underline{r}(s)$ with respect to s yields a unit tangent vector $\underline{t}(s)$.   This vector gives the "direction" of the curve at s; the rate of change in this direction with respect to s , that is $\underline{t}'(s)$, is a vector whose length is the *curvature* κ of the curve at s .

The solution to our problem can now be simply stated.   If our plane curve $\bar{C}$ is a feasible bending curve with the usual description $y = f(x)$ and curvature $\bar{\kappa}$, then the resulting space curve has curvature κ given by

$$\kappa^2 = \bar{\kappa}^2 + \lambda^4/\alpha^2 \tag{1}$$

where $\lambda^2 = [1 + (dy/dx)^2]^{-1}$ and α is the radius of the cylinder.   Also one can show that the angle θ between the cylinder and the collar at any point is given by

$$\cos\theta = 1 - 2\bar{\kappa}^2/\kappa^2$$

$$= \frac{\lambda^4 - \alpha^2\bar{\kappa}^2}{\lambda^4 + \alpha^2\bar{\kappa}^2} . \tag{2}$$

Now it is clear that θ must be a continuous function of position around the space curve. Otherwise the collar would certainly be torn.   Thus equation (2) implies that both λ and $\bar{\kappa}$ must in practice be continuous.   Continuity of λ is ensured if $\bar{C}$ has continuous gradient, itself a reasonable smoothness condition.   But the necessary continuity of $\bar{\kappa}$ is most interesting.   It means that the plane curve $\bar{C}$ must have continuous curvature.   This explains why the company had no success in constructing a bending curve from segments of circles. Even though circles can be made to join with matching gradients, there is always a step discontinuity in curvature where they join.

This restriction on $\bar{\kappa}$ turns out to be the crucial requirement in finding appropriate bending curves. In fact many bending curves will do the job. One family of curves proved particularly useful for simple application. It allows one to construct feeders for varying cylinder radius α , varying height h of the curve and different angles at the top of the space curve between collar and cylinder. The family depends upon α , h and two numbers p and q which are complicated functions of the ratio h/α and the top angle. For $0 \leq x \leq 2\pi\alpha$,

$$y = h[1 + p\{1-\cosh q(1-x/\pi\alpha)\}] \ .$$

To use this form one needs a small table of p and q values. Such a table was prepared for the company.

For further details on the mathematics of this problem the reader is referred to Culpin [1].

REFERENCES

[1]   Culpin, D. A metal-bending problem, Math. Scientist, 5, 121-127, 1980.

[2]   Weatherburn, C.E. Differential Geometry of Three Dimensions, Vol. I. Cambridge University Press, 1927.

FABRICATION OF OPTICAL FIBRES

R.B. Calligaro and D.N. Payne,
Electronics Department, Southampton University,

AND

R.S. Anderssen and B.A. Ellem,
Division of Mathematics and Statistics,
CSIRO, Canberra.

ABSTRACT

The application examined in this paper is the development of quality control techniques for optical fibre fabrication. It is an example of an application where non-destructive analysis and testing must be applied. The resulting mathematical problem reflects this as it reduces to solving an Abel integral equation which involves an indeterminacy.

1. INTRODUCTION

An optical fibre consists of a solid, but flexible, thin cylinder of silica glass with an axial symmetric refractive index. Fibres are constructed so that the index of the core is higher than that of the surrounding layers. In theory, light introduced into the core will be transmitted axially by total internal reflection. In practice, however, quite precise specifications, in terms of the purity of materials used and the consistency of the axial symmetric structure, must be met before fibres with high quality transmission characteristics can be fabricated. When fabricating a fibre, the major goals are to minimize the fibre attenuation and pulse dispersion. In addition, linear polarization fibres require the input state of polarization to remain unchanged and the distortion of any other special characteristics such as pulse dispersion to be minimized.

Optical fibres with high performance characteristics represent the basis for a future revolution in communications technology and sensor design. Though the revolution has in fact commenced, it will only attain fruition when fabrication procedures have been developed which guarantee that fibres with very precise specifications can be mass produced.

Thus, for fabrication purposes, it is necessary to have methods for determining the refractive index or equivalent profiles of optical fibres to enable prediction of their characteristics.   For example, such characteristics can often be derived non-destructively from a transverse optical examination of the preform from which the fibre is eventually pulled.

In fact, because of the photo-elastic effect, which states that the refractive index of a material depends on the distribution of elastic stresses within it, information about the refractive index or equivalent structure of the preform can *only* be obtained using non-destructive testing and analysis.   One possibility would be to make observations which could then be used to determine the refractive index profile directly.   However, since the refractive index is directly related to the elastic stresses via the photo-elastic effect, another possibility would be to make observations of the optical retardation because they can be used to determine the distribution of such stresses.

It is this possibility which we examine in this paper.   In fact, we examine the use of optical retardation data as a basis for quality control in preform fabrication.   In particular the usefulness of the elastic stress distributions for assessing the axial structure of the preform is explored.

Initially, in 2, additional background material about optical fibre fabrication and applications is developed.   After specifying the relationships between the cylindrical stress components and the thermal expansion coefficients in 3, the Abel equation formulation for the optical retardation in terms of the thermal expansion coefficient is derived in 4.   The solution of the formulation in 3 and 4 for the stresses is discussed in 5, while the experimental procedure for determining the optical retardation data is discussed in 6.   The usefulness of the axial and radial stresses as quality control measures for the fabrication of optical fibre preforms is then discussed in 7.

2. BACKGROUND AND APPLICATIONS

The idea of communications based on optical rather than electrical signals is not new, but refinements in technology, which have made it a possibility, are.   The two technological advances which have made the use of optical fibres a promising economic alternative in communications technology and sensor design are : (i) the invention and commercialization of lasers; and (ii) the development of high purity material processing.   The latter is being used for the construction of low loss fibres, while the former is being used to introduce light signals of high intensity and at a specific wavelength into the core.

A typical optical fibre consists of a fine strand of high purity silica glass with a typical diameter of 100 μm .   In its manufacture, in order to minimize contact with impurities successive layers of doped silica are deposited within a closed rotating silicate tube (*the substrate*) to form *the cladding region* which will eventually surround the core.   After the

construction of a suitably large number of layers, the process is terminated with a doped
silica essentially different from that used for the cladding, since it eventually forms *the
higher refractive index core*.   The tube is then collapsed to form a solid rod (the preform)
which thereby consists of the axially symmetric substrate and cladding layers around a central
core (cf. Figure 1).   The preform is subsequently heated and pulled to form the fibre.

FIGURE 1 : Radial variation of axial
refractive index $n_z(r)$ in a typical
optical fibre preform. The light intensity
is proportional to the axial stress $\sigma_z(r)$
which is proportional to the axial
refractive index $n_z(r)$. The higher
refractive index core, down which the light
is transmitted by total internal reflection, is clearly visible.

Because it has been found that, except possibly for minor effects of the pulling on the
core, the final structure of a fibre can be related to the initial structure of its preform,
the properties of fibres are usually interpreted in terms of the corresponding properties of
their preforms.

The optical transmission capabilities of a fibre are due directly to the higher refractive
index and geometry of the central core, since optical energy will be confined there by total
internal reflection.   For example, when the diameter of the core is kept suitably small, a
fibre can be obtained which only supports a single mode of transmission.   The advantages of
such *single mode fibres* is the very wide bandwidths and very high transmission capabilities
which they can support.

In fact, single mode fibres which maintain the input state of polarization throughout
their length are of great interest in instrumentation and telecommunications.   The former
include the Faraday effect ammeter {Smith   [13]   and Rashleigh and Ulrich [11]}   and interferometric devices such as the fibre gyroscope {Vali and Shorthill   [17]}   and strain and
pressure gauges {Butter and Hocker [4] and Hughes and Jarzynski [8]}.   The sensitivity
and stability of the Faraday effect ammeter is improved and, in the interferometric devices,
noise and drift are reduced when linear polarization fibres are used.   In telecommunications,
single mode fibres with high information capacity are suitable for interfacing with integrated
optical devices, especially when the properties of the devices are polarization sensitive
{Steinberg and Giallorenzi [14]}.

Two techniques are used to produce fibres which maintain the input state of polarization.
Either the stress mismatch between core and cladding along with the ellipticity of the fibre's
core are minimized {Norman, Payne, Adams and Smith   [9]} ;   or one of these properties is
strongly accentuated {Dyott, Cozens and Morris   [6] ,   and Stolen, Ramaswamy, Kaiser and Pleibel
[15]}.

The effect of core ellipticity {Adams, Payne and Raydal [1]} and mechanical distortions {Rashleigh and Ulrich [11] , and Hartog, Conduit and Payne [7]}   are well documented. However, the effect of the elastic stress, which is present in the preform by virtue of its composition of glasses cf unequal thermal expansions formed at elevated temperatures, has only been explained qualitatively.   As a fabricated preform (or fibre) cools below its softening (fictive) temperature, elastic stresses are set up as a direct consequence of the differences in thermal expansions between the different forms of silica making up the preforms axial structure.   On the other hand, the refractive index structure of the preform is related directly, via the photo-elastic effect, to the distribution of the elastic stresses within it.

Thus, for various reasons associated with fabrication, methods are required for determining the refractive index profiles and stress patterns in preforms and fibres.   However, as a direct consequence of the photo-elastic effect, non-destructive testing and analysis must be used.   The obvious way to do this is to exploit the transparency of the preforms and fibres by a transverse optical examination.   One possibility would be to make observations which could then be used to determine the refractive index profile directly.   Because the stress patterns are intrinsically important for the reasons listed above and the refractive index is related to them via the photo-elastic effect, we explore the use of optical retardation data since it can be used to determine the stresses.

In particular, we examine the usefulness of the elastic stress distributions as quality control indicators for assessing the axial structure of a preform.

3. THE RELATIONSHIPS BETWEEN THE ELASTIC STRESSES AND THE THERMAL EXPANSION COEFFICIENT

The addition of depont to the silica glass during the manufacture of preforms changes its thermal expansion coefficient and hence the residual stress pattern it takes as it solidifies. The nature of these changes is directly related to the type and quantity of depont introduced. Assuming the preform cools to form an axial symmetric structure, the form of the residual stresses set up depends on (i) the radial variation in the thermal expansion coefficient, $\alpha(r)$, within the preform, (ii) the size of the preform, and (iii) the difference ΔT , between room temperature and the fictive (or softening) temperature, at which cooling started. In fact, for an axial symmetric structure which has formed by cooling, the relationships between the internal stresses and the thermal expansion coefficient $\alpha(r)$ are given by (cf. Timoshenko and Goodier [16])

$$\sigma_r(r) = K \left\{ \frac{M_1}{R^2} - \frac{1}{r^2} \int_0^r \tau\alpha(\tau)\,d\tau \right\} , \tag{1}$$

$$\sigma_\theta(r) = \sigma_z(r) - \sigma_r(r) , \tag{2}$$

$$\sigma_z(r) = K \left\{ \frac{2M_1}{R^2} - \alpha(r) \right\} , \tag{3}$$

where σ_r , σ_θ and σ_z denote the stress components in the cylindrical coordinate directions r , θ and z ,

$$K = E.\Delta T/(1-\sigma) \ , \tag{4}$$

and

$$M_1 = \int_0^R r\alpha(r)\,dr \ , \tag{5}$$

with E denoting Young's modulus, σ Poisson's ratio, and R the outer radius of the preform.

4. THE ABEL EQUATION FORMULATION FOR OPTICAL RETARDATION

Because the photo-elastic effect necessitates non-destructive techniques being applied to determine the refractive index profile or stress patterns in a preform, it not only plays a crucial role physically, but also mathematically. In fact, it is the starting point for deriving the Abel equation formulation which defines optical retardation in terms of the thermal expansion coefficient $\alpha(r)$.

For an isotropic optical medium in which the components of the dielectric tensor are related linearly to the stress components, the photo-elastic effect can be defined formally as (cf. Born and Wolf [5], §14.5.1, equation (10))

$$n_x - n_z = C(\sigma_x - \sigma_z) \ , \tag{6}$$

where n_x and n_z denote the refractive indices in the x and z directions shown in Figure 2, σ_x and σ_z the corresponding stress components, and C the photo-elastic constant of the medium.

FIGURE 2

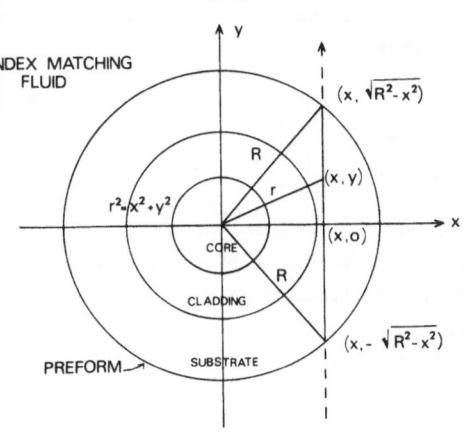

FIGURE 2 : Geometry of the preform and the path of the transverse polarized light.

Ne-He laser beam

(axial direction Z perpendicular to plane of Figure)

The photo-elastic effect is often exploited by measuring the optical retardation experienced by light passing through the preform normal to its axis. In fact, when such a beam of light is incident on the preform, it splits into two linearly polarized waves which are mutually orthogonal in the plane of polarization. Now, the phase difference δ_* between a normally-incident monochromatic plane wave of wave length λ (in vacuum)

$$a \cos \omega t$$

and the corresponding transmitted wave

$$a \cos(\omega t + \delta_*) ,$$

which propagates with wave length λ_* and velocity v_* through a medium of thickness h, satisfies (cf. Born and Wolf [5], §14.4.2)

$$\delta_* = \frac{2\pi}{\lambda_*} h = \frac{2\pi}{\lambda} n_* h ,$$

since, by definition, $n_* = c/v_*$ and $2\pi/\lambda_* = \omega/v_*$ (cf. Born and Wolf [5], §§1.2 and 1.3.3) where n_* denotes the refractive index of the medium and c the velocity of light in vacuum.

Thus, in terms of the geometry of Figure 2, the phase difference $d\delta$ between these two waves over a path length $d\ell$ becomes, on using the above result,

$$d\delta = \frac{2\pi d\ell}{\lambda}(n_x - n_z) .$$

On applying (6), which characterize the photo-elastic effect, this last result reduces to

$$d\delta = \frac{2\pi C d\ell}{\lambda}(\sigma_x - \sigma_z) . \tag{7}$$

Defining the *optical retardation* of the light as the total phase difference between these two waves as the light exits from the preform, it follows that

$$\delta(x) = \frac{2\pi C}{\lambda} \int_{-\sqrt{R^2-x^2}}^{\sqrt{R^2-x^2}} \{\sigma_x(y) - \sigma_z(y)\} dy , \tag{8}$$

where it has been assumed that the photo-elastic constant C remains constant along any ray path. In fact, when the doping of the silica is at very low levels, the photo-elastic constant is essentially that of pure silica glass. If, in addition, it is assumed that no shear stresses are induced by the mechanical and thermal processes of preform fabrication, and further, that the outer boundary of the preform is free from stresses, it follows that {cf. Poritsky [10], p.411}

$$\int_{-\sqrt{R^2-x^2}}^{\sqrt{R^2-x^2}} \sigma_x(y) dy \equiv 0 .$$

After applying this last result, (8) yields, when combined with (3),

$$\delta(x) = D \left\{ \int_0^{\sqrt{R^2-x^2}} \alpha(y)\,dy - \frac{2M_1}{R^2} (R^2-x^2)^{\frac{1}{2}} \right\} \tag{9}$$

with

$$D = 4\pi CK/\lambda . \tag{10}$$

Introducing the substitution $r^2 = x^2 + y^2$ (cf. Figure 2), (9) transforms to the following Abel integral equation

$$f(x) = \int_x^R \frac{r\alpha(r)}{\sqrt{r^2-x^2}} \, dr \tag{11}$$

with

$$f(x) = \frac{\delta(x)}{D} + \frac{2M_1}{R^2}(R^2-x^2)^{\frac{1}{2}} . \tag{12}$$

Thus, the determination of the stress profiles (1), (2) and (3) is reduced to solving (11) and (12).

5. SOLUTION OF THE ABEL EQUATION FORMULATION FOR THE RADIAL STRESS

The Abel equation formulation (11), (12) and (5) is non-standard in that it contains an indeterminacy. The unknown $\alpha(r)$ can be defined in terms of $f(x)$ by

$$\alpha(r) = -\frac{2}{\pi} \frac{1}{r} \frac{d}{dr} \left\{ \int_r^R \frac{xf(x)}{\sqrt{x^2-r^2}} \, dx \right\} , \tag{13}$$

which is one of the known inversion formulas for (11), but the form of $f(x)$ depends on M_1 which in turn depends on $\alpha(r)$.

Since the radial stress is defined as an indefinite linear functional on the thermal expansion coefficient $\alpha(r)$, the inversion formula (13) can be used in conjunction with (11) to redefine the stress as an indefinite linear functional on $f(x)$ {cf. Anderssen (1977), (1980), and Golberg (1979)}. In this way, Anderssen and Calligaro (1980) have shown that the radial stress can be evaluated as the following indefinite linear functional on the retardation data $\delta(x)$

$$\sigma_r(r) = \frac{\lambda}{2C\pi^2 r^2} \int_r^R \frac{x\delta(x)}{\sqrt{x^2-r^2}} \, dx . \tag{14}$$

The advantage of this formula is that it does not involve M_1 ; i.e., the indeterminacy has been removed.

They also showed that the indeterminacy associated with the evaluation of the circular and axial stresses $\sigma_\theta(r)$ and $\sigma_z(r)$ is removed if they are calculated as follows

$$\sigma_z(r) = -\frac{2}{\pi D}\frac{1}{r}\frac{d}{dr}\left\{\int_r^R \frac{x\delta(x)}{\sqrt{x^2-r^2}}\right\} = -\frac{1}{r}\frac{d}{dr}\{r^2\sigma_r(r)\}, \tag{15}$$

$$\sigma_\theta(r) = \sigma_z(r) - \sigma_r(r) = \frac{d}{dr}(r\sigma_r(r)) \tag{16}$$

6. THE OPTICAL RETARDATION DATA

The experimental set-up is shown in Figure 3. The preform was immersed in index matching fluid and illuminated transverse to its axial direction with a He-Ne laser with $\lambda = 0.6328$ μm . Before entering the preform, the laser beam was passed through a half-wave plate and polarizer. On exit, the beam was passed through a quarter-wave plate and analyser before entering a photodetector with an aperture of 50 μm .

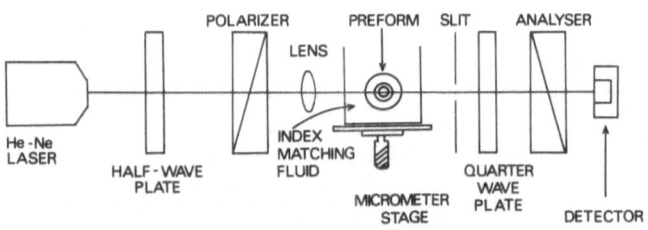

FIGURE 3 : Schematic diagram of experimental setup.

The half-wave plate simply rotates the polarization state of the input light. It is used to rotate the laser light into a favourable position for input through the polarizer. The polarizer only transmits a linearly polarized component of the input laser light. We assume that this linear polarization (in the plane perpendicular to its direction of propagation which from Figure 2, is the y-direction) has the representation (cf. Figure 4)

$$z' = 0$$

$$x' = \sqrt{2}\ a \sin \omega t , \quad a = const.,$$

where ω denotes the (angular) frequency of the coherent laser light. The axis of the polarizer is set at an angle θ to the axial direction (the z-direction from Figure 2) of the preform (and parallel to the axis of the quarter-wave plate). The polarized light therefore falls with normal incidence on the specimen.

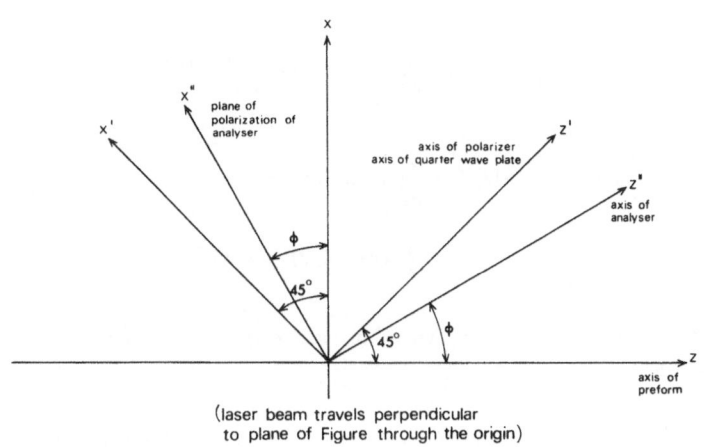

(laser beam travels perpendicular
to plane of Figure through the origin)

FIGURE 4 : Relative positions of the directions of polarization associated with
the key components in the experimental setup.

Due to the stresses in the preform and the photo-elastic effect, the refractive index
throughout the preform will have an axially symmetric structure and therefore have a two-
dimensional structure in the plane of polarization of the incoming laser light.   Thus, on
entering the specimen, the polarized light will be split into two components.   One is in the
z-direction, the other in the x ; viz.

$$z = z' \cos \theta = \sqrt{2} \ a \cos \theta \sin \omega t \ ,$$

$$x = x' \sin \theta = \sqrt{2} \ a \sin \theta \sin \omega t \ .$$

As the velocity of light in a medium depends on its refractive index, it follows that
these two components will travel through the preform with different speeds exiting with a
relative phase (time) delay of δ ; i.e.

$$z = \sqrt{2} \ a \cos \theta \sin (\omega t \pm \delta) \ ,$$

$$x = \sqrt{2} \ a \sin \theta \sin (\omega t) \ .$$

This will be circularly polarized if $\theta = \pi/4$.

The quarter-wave plate is placed between the preform and analyser to ensure that each
of the components of the elliptically polarized light examined by the analyser contains a factor
involving the phase delay δ .   When the axis of the quarter-wave plate is set at an angle of

$\pi/4$ to the axial direction of the preform (and parallel to that of the polarizer), the quarter wave plate resolves the light, as it enters from the preform, into the following components

$$z' = (z-x)/\sqrt{2} = \frac{a}{\sqrt{2}} \, [\sin \, (\omega t \pm \delta) - \sin \, \omega t]$$

$$= \frac{2a}{\sqrt{2}} \, \sin \, (\pm\delta/2)\cos \, (\omega t \pm \delta/2) \; ,$$

and

$$x' = (z+x)/\sqrt{2} = \frac{a}{\sqrt{2}} \, [\sin \, (\omega t \pm \delta) + \sin \, \omega t]$$

$$= \frac{2a}{\sqrt{2}} \, \cos \, (\delta/2) \, \sin \, (\omega t \pm \delta/2) \; .$$

If the z' - direction is the fast axis, then, on emerging from the quarter-wave plate, the components will have become

$$z' = \sqrt{2} \, a \, \sin \, (\pm\delta/2) \, \cos \, (\omega t \pm \delta/2 + \pi/2) = - \, \sqrt{2} \, a \, \sin \, (\pm\delta/2) \, \sin \, (\omega t \pm \delta/2) \; ,$$

$$x' = \sqrt{2} \, a \, \cos \, (\delta/2) \, \sin \, (\omega t \pm \delta/2) \; .$$

If the plane of polarization of the analyser is assumed to be in the x''-direction making an angle φ with the axis of the quarter-wave plate (and polarizer), then the component of the linearly polarized light leaving the analyser will be

$$z'' = 0$$

$$x'' = x' \sin \, \varphi + y' \cos \, \varphi$$

$$= \sqrt{2} \, a \, \sin \, (\omega t \pm \delta/2) \, [\cos \, (\delta/2) \, \cos \, \varphi - \sin \, (\pm\delta/2) \, \sin \, \varphi]$$

$$= \sqrt{2} \, a \, \cos \, (\varphi \pm \delta/2) \, \sin \, (\omega t \pm \delta/2) \; .$$

Since the intensity of light is proportional to the square of its amplitude, it follows that, as a function of φ , the intensity of the light entering the photodetector will be

$$I = I_0 \, \cos^2(\varphi \pm \delta/2) \; ,$$

where I_0 denotes the constant of proportionality. Thus, the intensity of the light reaching the photodetector will be "zero" if the analyser is adjusted so that φ satisfies

$$\varphi = \pm \, \delta/2 + \left(\frac{2k+1}{2}\right)\pi \; , \qquad k = 0,1,2,\ldots \; .$$

In this way, the relative retardation δ between the two polarized waves which leave the preform can be estimated as the angle $\delta/2$ that the analyser must be rotated, relative to the axis of the polarizer, to minimize the signal received at the detector.

Retardation profiles of the form shown in Figure 5 are obtained by determining the retardation angle δ as the preform is stepped across the laser beam on a micrometer stage. The accuracy of the stepping can be calculated to ±1 μm while that of the retardation angle is ±0.5° when δ is approximately constant and ±4° when the polarized light traverses the sharp gradients associated with the cladding.

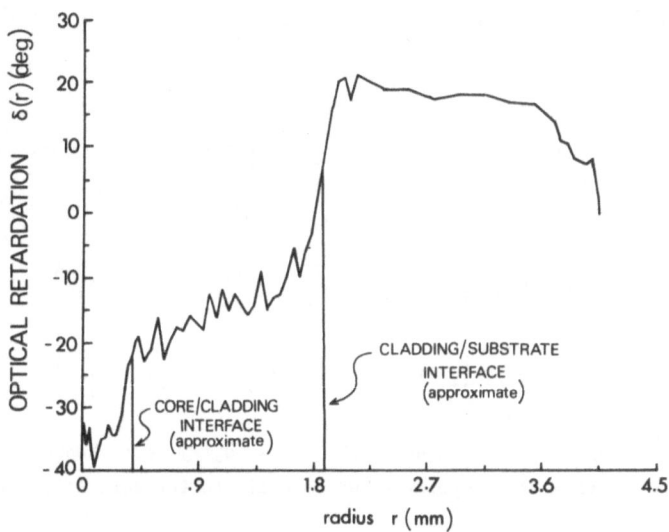

FIGURE 5 : A typical example of optical retardation data. The relative positions of the core/cladding and cladding/substrate interfaces are clearly visible in these data, but not the radial variation of $n_z(r)$.

7. RESULTS AND CONCLUSIONS

A typical optical retardation profile is shown in Figure 5. The relative (approximate) positions of the core/cladding and cladding/substrate interfaces are indicated since they are clearly visible. However, this profile contains no direct insight about the radial variation of the axial refractive index $n_z(r)'$. This is a key property since it is the structure of $n_z(r)$ which determines the transmission properties of the fibre pulled from the preform.

We therefore explore the use of the radial and axial stresses as a basis for gaining such insight.

In passing, we note that the retardation data satisfy the condition

$$\int_0^R \delta(r)\,dr = 0$$

which relates to the fact that the total axial stress across the preform must be zero (cf. Anderssen and Caligaro [3]).

For the retardation data shown in Figure 5, the radial stress $\sigma_r(r)$ was computed from (14) using the product trapezoidal rule; viz. on the grid

$$0 = x_0 < x_1 < \ldots < x_n = R , \quad \Delta_j = x_{j+1} - x_j , \quad j = 0,1,2,\ldots,n-1 ,$$

estimate

$$\sigma_r(r) = \int_r^R \frac{\theta(x)}{\sqrt{x^2-r^2}} \, dx , \qquad \theta(x) = x\delta(x) ,$$

as

$$\hat{\sigma}_r(r) = \sum_{j=k}^{n-1} \left\{ (\theta_j x_{j+1} - \theta_{j+1} x_j) \ell n \left[\{x_{j+1} + (x_{j+1}^2 - x_k^2)^{\frac{1}{2}}\} \cdot \{x_j + (x_j^2 - x_k^2)^{\frac{1}{2}}\}^{-1} \right] \Delta_j^{-1} \right.$$

$$\left. + (\theta_{j+1} - \theta_j)(x_{j+1}^2 - x_j^2) \left[\{(x_{j+1}^2 - x_k^2)^{\frac{1}{2}} + (x_j^2 - x_k^2)^{\frac{1}{2}}\} \Delta_j \right]^{-1} \right\} .$$

The results are shown in Figure 6a. The axial stress $\sigma_z(r)$ was then estimated from (15) using a weighted combination of mid-point rules to evaluate the derivative of $(r^2 \hat{\sigma}_r(r))$; viz.

$$\hat{\sigma}_z(x_k) = - \sum_{j=1}^{3} \frac{w_j \, (x_{k+j}^2 \hat{\sigma}_r(x_{k+j}) - x_{k-j}^2 \hat{\sigma}_r(x_{k-j}))}{x_k \, (x_{k+j} - x_{k-j})} ,$$

where the w_j denote weights. The retardation data were not measured on a sufficiently fine grid to warrant the application of a more sophisticated numerical differentiation procedure. The results for $\hat{\sigma}_z(r)$ are shown in Figure 6b.

In both Figures 6a and 6b, the approximate position of the core/substrate and substrate/ cladding interfaces are shown. It is clear that $\hat{\sigma}_z(r)$ yields a better qualitative picture of the structural components of the preform than $\hat{\sigma}_r(r)$; and therefore is the better indicator for quality control purposes. This is not too surprising as $\hat{\sigma}_z(r)$ monitors the behaviour of the axial refractive index $n_z(r)$, and as the fabrication process has been designed to produce a preform in which the variations of $n_z(r)$ as a function of r are quite sharply defined. The layering in the cladding is clearly visible in Figure 6b, but not in Figure 6a.

A transverse image (photograph) of the preform, from which the data of Figure 5 were obtained, is shown in Figure 6c. In this case, the experimental setup of Figure 3 was used with a mercury lamp replacing the He-Ne laser. The analyser was adjusted so that the background was black. In this way, the light intensity in the resulting photograph is directly proportional to the stress.

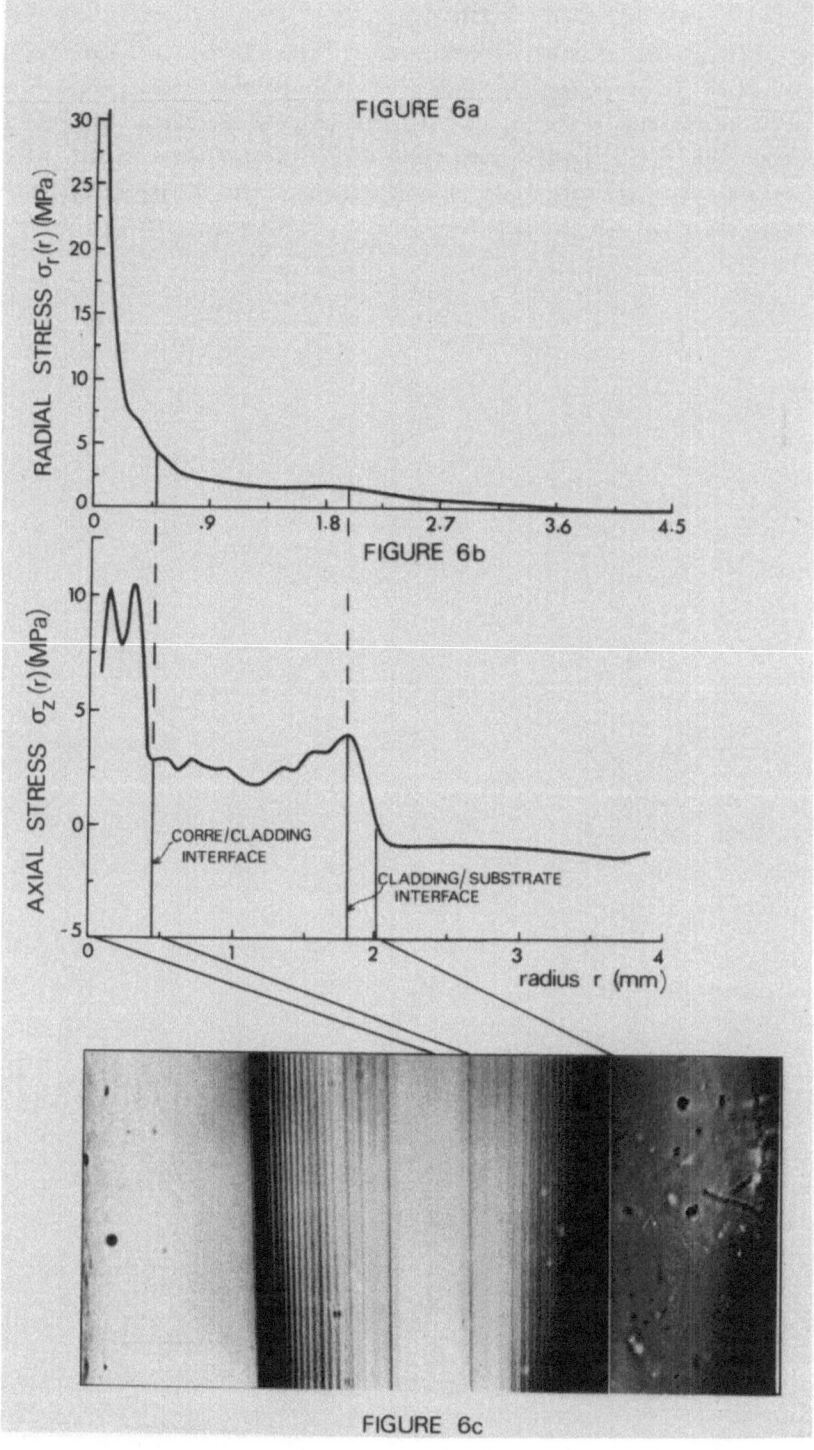

FIGURE 6a

FIGURE 6b

CORRE/CLADDING INTERFACE

CLADDING/SUBSTRATE INTERFACE

FIGURE 6c

FIGURE 6 : a. The radial variation of the radial stress; b. The radial variation of the axial stress; c. A transverse image of the preform in which light intensity is directly proportional to the stress. The key features in Figure 6b are identified with the corresponding features in the image.

It is clear that the key features in Figure 6b can be identified with corresponding features in the photograph of Figure 6c.   Because the retardation data were not recorded on a sufficiently fine grid, $\hat{\sigma}_z(r)$ does not have the resolution obtained in the photograph. However, the $\hat{\sigma}_z(r)$ profile does yield a sufficiently clear picture of the radial variation of itself and $n_z(r)$ for it to be used as a basis for a quality control indicator in optical preform and fibre fabrication.

REFERENCES

[1]   Adams, M.J., Payne, D.N. and Raydal, C.M. (1979) Birefringence in optical fibres with elliptical cross-section, *Elect. Letters* 15 (1979), 298-299.

[2]   Anderssen, R.S. (1976) Stable procedures for the inversion of Abel's equation, *JIMA* 17 (1976), 329-342.

[3]   Anderssen, R.S. and Calligaro, R.B. (1980) Nondestructive testing of optical fibre preforms, *J. Aust. Math. Soc., Series B* (in press)

[4]   Butter, C.D. and Hocker, G.B. (1978) Fibre optics strain gauge, *Appl. Opt.* 17 (1978), 2867-2869.

[5]   Born, M. and Wolf, E. (1975) *Principles of Optics* (th Edition) Pergamon Press, Oxford, 1975.

[6]   Dyott, R.B., Cozens, J.R. and Morris, D.G. (1979) Preservation of polarization in optical fibre wave guides with elliptical cores, *Elect. Letters* 15 (1979), 380-382.

[7]   Hartog, A.H., Conduit, A.J. and Payne, D.N. (1979) Variation of pulse delay with stress and temperature in jacketed and unjacketed optical fibres, *Opt. Quart. Elect.* 11 (1979), 265-273.

[8]   Hughes, R. and Jarzynski, J. (1980) Static pressure sensitivity amplification in interferometric fibre optic hydrophones, *Appl. Opt.* 19 (1980), 98-107.

[9]   Norman, S.R., Payne, D.N., Adams, M.J. and Smith A.M. (1979) Fabrication of single mode fibres exhibiting extremely low polarization birefringence, *Elect. Letters* 15 (1979), 309-311.

[10] Poritsky, H. (1934) Analysis of thermal stress in sealed cylinders and the effect of viscous flow during annealing, *Physics* 5 (1934), 406-411.

[11] Rashleigh, S.C. and Ulrich, R. (1979) Magneto-optic sensing with birefringent fibres, *Appl. Phys. Letters* 34 (1979), 768-770.

[12] Shibata, N., Junguji, K., Kawachi, M. and Edahiro, T. (1979) Nondestructive structure measurement of optical fibre preforms with photo-elestic effect, *Jap. J. Appl. Phys.* 18 (1979), 1267-1273.

[13] Smith, A.M. (1978) Polarization and magneto-optic properties of single mode optical fibre, *Appl. Opt.* 17 (1978), 52-56.

[14] Steinberg, R.A. and Giallorenzi, T.G. (1977) Design of integrated optic switches for uses in fibre data transmission systems, *IEEE J. of Quart. Elect.* QE-13 (1977), 122-128.

[15]   Stolen, R.H., Ramaswamy, V., Kaiser, P. and Pleibel, W. (1978) Linear polarization in birefringent single-mode fibres, *Appl. Phys. Letters* 33 (1978), 699-701.

[16]   Timoshenko, S. and Goodier, J.N. (1970) *Theory of Elasticity* (3rd Edition) McGraw-Hill, New York, 1970, p.

[17]   Vali, V. and Shorthill, R.W. (1976) Fibre ring interferometer gyro, *Appl. Opt.* 15 (1976), 1099-1100.

A SIMPLE MODEL FOR COIL INTERIOR TEMPERATURE
PREDICTION DURING BATCH ANNEALING

R.M. LEWIS,

Research and Technology Centre,
John Lysaght (Australia) Limited, Port Kembla.

1. INTRODUCTION

Thin steel sheet in the range 0.2 mm to 2.0 mm[1] thick is produced from 250 mm thick, 12 m long, cast slab[2] in three rolling operations [3]. The first two operations, reverse roughing, which reduces the slab to a 25 mm thick, 120 m long transfer bar, and hot rolling, which results in a 2.5 mm thick, 1200 m long strip which is wound into a coil, are performed at elevated temperatures, usually above 900°C. The final reduction operation, cold rolling is performed at low temperature, below 100°C, to obtain a better surface finish than achievable by hot rolling. Each thickness reduction deforms the crystal structure of the steel into elongated grains, reducing the formability of the material. The crystal structure and formability is restored provided there is sufficient thermal energy in the steel to promote recrystallization - during hot rolling recrystallization occurs after each stage of reduction but during cold rolling it does not, so that after cold rolling a special heat treatment process, annealing, is required [4,5].

Two distinct methods of annealing, continuous and batch, are in use. With continuous annealing furnaces, the steel coil is unwound at the entry end and fed through the furnace as a continuous strip, being heated and cooled in different furnace sections over a period of several minutes before being either recoiled at the exit or fed directly into the next processor, such as the metal bath of a continuous galvanizing line. By contrast, in a batch annealing furnace (see Figure 1), the tightly wound steel coils are stacked end upon end and heated and cooled over a period of several days. Owing to their lower capital cost and energy usage and despite their slowness and non-integratability with other continuous processors, there has recently been a revival of interest in batch annealing installations, and consequently a renewal of research effort aimed at improving furnace efficiency and product uniformity.

---

1. All figures are typical.
2. Such slabs are obtained directly from continuous casting or by primary rolling ingot castings.

The research program of John Lysaght (Australia) has centred on the development of a comprehensively detailed mathematical model of batch annealing furnaces, since this approach is far less costly than the conduct of numerous plant temperature measurement trials for the development of an empirical model (a small number of trials were required to confirm model predictions and determine inaccurately known furnace parameters). The model, described in [2 is essentially a collection of finite difference equations representing heat transfer by radiation and convection from the combustion gas to the coil surface via the furnace wall, cover, protective annealing gas stream and convector plates (see Figure 1; for a furnace with three coils stacked on top of each other) and heat diffusion within the coils. The uses of this detailed model included furnace design analyses which have led to improved heat transfer and reduced annealing times, annealing cycle development which has enabled the annealing of high strength steels and the development of the simple model described in this paper.

BATCH ANNEALING FURNACE CROSS-SECTION

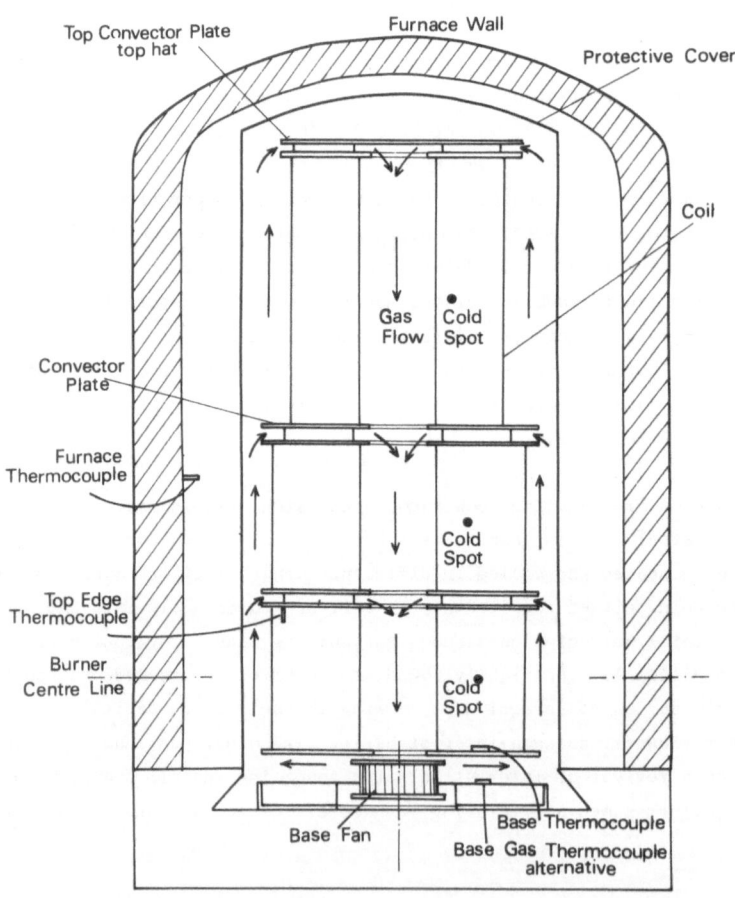

FIGURE 1.

The basic concepts for on-line control of the furnaces have remained substantially unchanged for three decades largely because of the inability to measure the temperatures inside coils.   The development of accurate predictive models combined with the advent of relatively cheap, powerful micro-processors has made possible both the on-line calculation of these internal temperatures from measured external temperatures and the use of this information to control the furnaces.   On-line use of the detailed model referred to above is precluded by the size (60k bytes) and speed of the simulation package.   A simple model for on-line use has therefore been constructed, using information and experience gained from the detailed simulations.

2. MODEL DESIGN OBJECTIVES AND METHODS

The annealing of standard drawing quality steel sheet requires the material to be heated to 620°C whilst not exceeding the critical temperature for the formation of an undesirable pearlite crystal structure, 720°C.   The present annealing cycle for this material is specified as follows :

Model 1 :   furnace on full fire until the furnace thermocouple (TC1, see Figure 1) reaches 850°C.

Model 2 :   hold TC1 at 850°C until the top edge thermocouple (TC2) reaches 690°C.

Model 3 :   hold TC2 at 690°C until the base thermocouple (TC3) reaches 620°C.

Model 4 :   hold TC2 at 690°C for a further 6 hours (soak period).

Model 5 :   remove furnace and allow natural cooling until TC3 reaches 550°C.

Model 6 :   use forced cooling until TC3 reaches 130°C (forced air cooling) or 100°C (water cooling).

Notes   (i)   The top edge temperature is usually within 30°C of the highest temperature recorded in any of the coils and is used to avoid overheating.

(ii)   The base thermocouple measurement can be replaced by the more reliable base gas temperature measurement with a suitable adjustment of the soak time in mode 4 and the temperatures in modes 5 and 6.

(iii) Forced air cooling is accomplished using a cooling tower with fans directing air at the cover whilst water cooling is employed on bases fitted with heat exchanger coils in the gas stream around the fan, by passing water through the coils.

Plant trials and detailed simulations confirm that the above cycle results in coil minimum temperatures in the vicinity of 620°C after the soak period.   However the use of a fixed duration soak period naturally results in the variation of minimum temperature from charge to charge, hence some variation in the mechanical properties of the steel produced. Further, the safety margin built into the soak time to avoid underannealing results in unnecessarily long cycle times on some charges and therefore a loss of production.

Clearly, the annealing operation can be made more efficient with the knowledge of coil minimum temperatures.   Direct measurement of these temperatures in normal production is not possible due to the time and consequent cost penalty associated with the insertion of thermocouples into coils.   At present the only practical method for obtaining them is by on-line calculation from measured external coil temperatures.

The flow of heat in each coil is described by the non-linear axisymmetric heat conduction equation :

$$\frac{\partial(\gamma cT)}{\partial t} = \frac{1}{r}\frac{\partial}{\partial r}\left[rk_r\frac{\partial T}{\partial r}\right] + \frac{\partial}{\partial z}\left[k_z\frac{\partial T}{\partial z}\right] \tag{1}$$

with boundary conditions :

$$k_n\frac{\partial T}{\partial n} = \sum_i q_{n_i} \tag{2}$$

where   n   is the direction normal to a surface.

The axial and radial conductivities k_z and k_r are strongly temperature dependent and in addition, because of differential coil wrap expansion, k_r depends upon the gradient $\partial T/\partial r$,   as discussed in [2].   The right hand side of equation (2) is the sum of all radiative and convective heat flows into the coil surface under consideration.

The selection of an approximate method for the on-line solution of (1) must be based upon the availability of temperature measurements from which the heat flows into the coils can be calculated.   For the calculation of heat flows into coil outer wraps, annealing gas and cover temperatures opposite each coil are required.   For the determination of such information thermocouples would have to be suspended in the gas stream and attached to the cover every time a charge is fired.   This is precluded by its practical difficulty and we must therefore do without the heat flows into outer wraps.   Fortunately, trial measurements and detailed simulations show that temperature distributions in coils have a high degree of symmetry about coil cold spots - the locations of coil temperature minima (cf. Figure 1). Moreover, it is observed that cold spot positions (lying in the range from $\frac{1}{4}$ to $\frac{1}{2}$ the radial distance from bore to outer wrap and approximately mid width position axially) are only strongly dependent upon coil positions in the furnace and not upon other factors.   Thus, assuming that the temperature is symmetric about the cold spot, it follows that (see Figure 2).

$$\left.\frac{\partial T}{\partial r}\right|_{r=r_c} = 0 \ , \qquad \left.\frac{\partial T}{\partial z}\right|_{z=0} = 0$$

$$r_c = \text{cold spot radius}$$

An on-line solution method can therefore be constructed once heat input at the bore and at the upper or lower ends of the coil have been estimated.

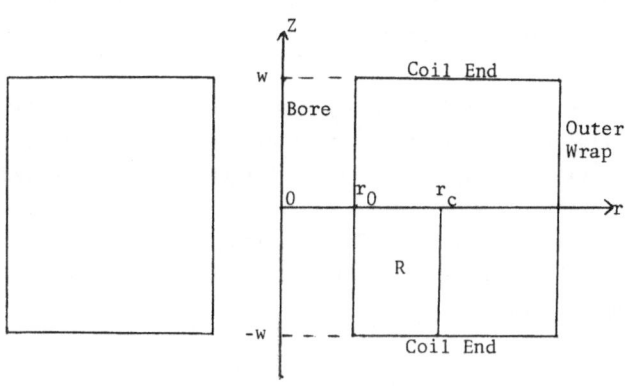

r_c = cold spot radius

w = coil half width

FIGURE 2 : Coil Cross Section

Heat input to a coil bore is determined (see Appendix 2 for further details) by bore gas temperature while heat input to a coil end is determined by convector gas temperature, indirectly via the convector plate.   Trial and simulation observations reveal that both these gas temperatures are approximately equal to the base gas temperature with maximum differences of 50°C early in a cycle rapidly reducing to less than 20°C.   Since the gas-to-coil and gas-to-convector convection coefficients are not known to better than within 10%, it appears reasonable to base all heat input calculations on the already existing base gas thermocouple temperature measurement.   This restriction of heat input calculations to the use of an existing temperature measurement greatly enhances the attractiveness of the resulting model for on-line control.   (Note : the base gas thermocouple which is permanently located in the furnace base cannot be damaged during charge building and hence is extremely reliable.)

Given the degree of approximation in the boundary conditions it would be pointless to solve the heat equation using a method accurate to better than about 5°C.   The observed symmetric temperature profiles can be approximated with this accuracy by quartic functions, hence we seek a solution in the functional form :

$$T = (a+bz^2+cz^4)(d+e\rho^2+f\rho^4) \tag{3}$$

where $\rho = r-r_c$, r_c = cold spot radius and a-f are time dependent.   In the context of variational techniques, equation (3) specifies a test function.   This approximation ultimately yields a 4th order dynamic system representation of equations (1) and (2) - its accuracy is far greater than a 9 point finite difference approximation which would lead to a dynamic system of the same order, because the quartic representation accurately reflects the large temperature gradients present near coil boundaries.

As already noted, because of the temperature dependence of the conductivities k_z and k_r , equation (1) is non-linear.   The conductivities are non-linear functions of temperature hence to avoid using look-up tables on-line, equation (1) is solved in linearized form with all coefficients evaluated at a nominal temperature of 450°C (the radial conductivity is evaluated at a nominal gradient $\partial T/\partial r$) .   This results in a simple on-line model with a small number of parameters which can be adjusted to give the best agrement between predicted and recorded cold-spot temperatures.

Note : The above description of modelling objectives and assumptions may have given the reader the impression that the whole exercise is strongly dependent upon the existence of the detailed simulation, itself the subject of a comprehensive modelling exercise. However, while the detailed simulation was used extensively to check the validity of assumptions made, all these checks could have been performed using plant trial data.

3. AN APPROXIMATE SOLUTION TO THE HEAT EQUATION FOR A STEEL COIL

In this section we derive equations whose solution approximates the time evolution of temperatures in a steel coil during annealing.   The equations are derived from an approximate solution to the heat equation of quartic functional form, equation (3).   The properties of symmetric quartic functions are summarized in Appendix 1.

The heat equation to be solved is a linearized version of equation (1)

$$\frac{1}{a} \frac{\partial T}{\partial t} = \frac{K}{r} \frac{\partial}{\partial r} \left[r \frac{\partial T}{\partial r} \right] + \frac{\partial^2 T}{\partial z^2} \tag{4}$$

where $a = k_z/\gamma c$ is the diffusivity of steel and $K = k_r/k_z$ is the relative radial conductivity.   We seek an approximate solution over the region R indicated in Figure 2. An enlarged picture of R is shown in Figure 3.   We approximate equation (4) at the coil interior points A , B , C and D shown in Figure 3 using expressions for $\frac{1}{r} \frac{\partial}{\partial r} \left(r \frac{\partial T}{\partial r} \right)$ and $\frac{\partial^2 T}{\partial z^2}$ derived from equations (A4) and (A5) of Appendix 1.   For example at point A :

$$\frac{\partial^2 T}{\partial z^2} \doteq \frac{1}{6w_1^2} \left[-15T(A) + 16T(C) - T(H) \right] \tag{5a}$$

$$\frac{1}{r} \frac{\partial}{\partial r} \left[r \frac{\partial T}{\partial r} \right] \doteq \frac{1}{6\rho^2} \left[-15T(A) + 16T(B) - T(E) \right] \tag{5b}$$

and at point D ;

$$\frac{\partial^2 T}{\partial z^2} \doteq \frac{1}{6w_1^2} \left[- 8T(D) + 3T(B) + 5T(G) \right] \tag{5c}$$

$$\frac{1}{r} \frac{\partial}{\partial r} \left[r \frac{\partial T}{\partial r} \right] \doteq \frac{1}{6\rho^2} \left[-8T(D) + 3T(C) + 5T(F) \right] + \frac{1}{6\rho r_1} \left[- 8T(D) + 9T(C) - T(F) \right] \tag{5d}$$

ρ , r_1 and w_1 are defined in Figure 3.   At A , B , C and D the left hand side of (4) becomes $\frac{1}{a} \frac{dT(A)}{dt}$, $\frac{1}{a} \frac{dT(B)}{dt}$ etc.

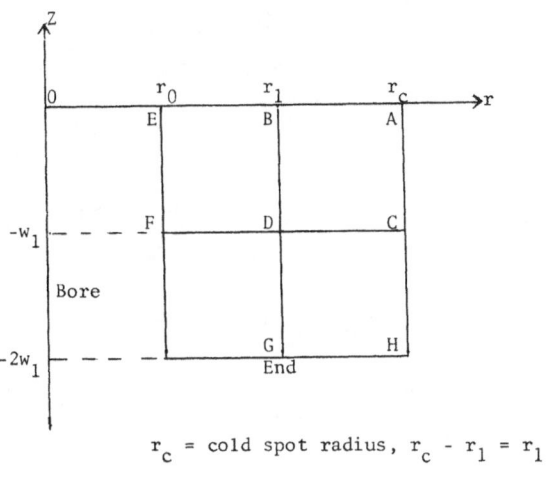

r_c = cold spot radius, $r_c - r_1 = r_1 - r_0 = \rho$

w_1 = coil quarter width

FIGURE 3 : Enlargement of Region R.

This approximation to equation (4) results in four simultaneous ordinary differential equations for T(A) , T(B) , T(C) and T(D) .   The inputs to this system are the coil boundary temperatures T(E) , T(F) , T(G) and T(H) - however, using the boundary condition (equation (2)) the boundary temperatures can be expressed in terms of internal temperatures and the measured gas stream temperature TG .   In Appendix 2 it is shown that all the boundary conditions can be expressed as

$$k_n \left. \frac{\partial T}{\partial n} \right|_X = - H_X \, (TG - T(X)) \tag{6}$$

where X = E , F , G or H .   Equation (A4) is then used to express $\left. \frac{\partial T}{\partial n} \right|_X$ in terms of T(X) , T(A) , T(B) , T(C) and T(D) , for example at X = E equation (6) becomes :

TABLE 1

COEFFICIENTS IN EQUATION 9

$$a_{11} = -\frac{5}{2a}\left(\frac{1}{w_1^2} + \frac{K}{\rho^2}\right) + \frac{3}{2a}\left(\frac{C_H}{w_1^2} + \frac{C_E K}{\rho^2}\right)$$

$$a_{12} = \frac{8K}{3a\rho^2}(1 - C_E)$$

$$a_{13} = \frac{8}{3aw_1^2}(1 - C_H)$$

$$a_{14} = 0$$

$$a_{21} = \frac{K}{2a\rho}\left[\frac{1}{\rho} + \frac{3}{r_1} - 3C_E\left(\frac{5}{\rho} - \frac{1}{r_1}\right)\right]$$

$$a_{22} = \frac{1}{2aw_1^2}(3C_G - 5) + \frac{4K}{3a\rho}\left[2C_E\left(\frac{5}{\rho} - \frac{1}{r_1}\right) - \frac{1}{\rho} - \frac{1}{r_1}\right]$$

$$a_{23} = 0$$

$$a_{24} = \frac{8}{3aw_1^2}(1 - C_G)$$

$$a_{31} = \frac{1}{2aw_1^2}(1 - 15C_H)$$

$$a_{32} = 0$$

$$a_{33} = \frac{4}{3aw_1^2}(10C_H - 1) + \frac{K}{2a\rho}(3C_F - 5)$$

$$a_{34} = \frac{8K}{3a\rho^2}(1 - C_F)$$

$$a_{41} = 0$$

$$a_{42} = \frac{1}{2aw_1^2}(1 - 15C_G)$$

$$a_{43} = \frac{K}{2a\rho}\left[\frac{1}{\rho} + \frac{3}{r_1} - 3C_F\left(\frac{5}{\rho} - \frac{1}{r_1}\right)\right]$$

$$a_{44} = \frac{4}{3aw_1^2}(10C_G - 1) + \frac{4K}{3a\rho}\left[2C_F\left(\frac{5}{\rho} - \frac{1}{r_1}\right) - \frac{1}{\rho} - \frac{1}{r_1}\right]$$

$$b_1 = -\frac{1}{6a}\left(\frac{d_H}{w_1^2} - \frac{Kd_E}{\rho^2}\right)$$

$$b_2 = -\frac{1}{6a}\left(\frac{d_G}{w_1^2} + \frac{Kd_E}{\rho}\left(\frac{1}{r_1} - \frac{5}{\rho}\right)\right)$$

$$b_3 = \frac{1}{6a}\left(\frac{5d_H}{w_1^2} - \frac{Kd_F}{\rho^2}\right)$$

$$b_4 = \frac{1}{6a}\left(\frac{5d_G}{w_1^2} + \frac{Kd_F}{\rho}\left(\frac{5}{\rho} - \frac{1}{r_1}\right)\right)$$

$$T(E)\left[H_E + \frac{7k_r}{3\rho}\right] = H_E TG + \frac{k_r}{3\rho}[16T(B) - 9T(A)] \qquad (7)$$

Let $\qquad c_E = \left(\frac{k_r}{3\rho}\right) / \left(H_E + \frac{7k_r}{3\rho}\right) \qquad$ and $\qquad d_E = H_E / \left(H_E + \frac{7k_r}{3\rho}\right)$

then $\qquad T(E) = c_E (16T(B) - 9T(A)) + d_E TG \qquad (8)$

Similar expressions can be obtained for c_X , d_X and $T(X)$ when $X = F$, G , or H .

Substitution of (8) (and the similar expressions for $T(F)$, $T(G)$ and $T(H)$) into equations (5) (and the similar equations for points B and C) results in the following dynamic system approximation to equation (4) and its boundary conditions.

$$
\begin{aligned}
dT(A)/dt &= a_{11}T(A) + a_{12}T(B) + a_{13}T(C) + a_{14}T(D) + b_1 TG \\
dT(B)/dt &= a_{21}T(A) + a_{22}T(B) + a_{23}T(C) + a_{24}T(D) + b_2 TG \\
dT(C)/dt &= a_{31}T(A) + a_{32}T(B) + a_{33}T(C) + a_{34}T(D) + b_3 TG \\
dT(D)/dt &= a_{41}T(A) + a_{42}T(B) + a_{43}T(C) + a_{44}T(D) + b_4 TG
\end{aligned}
\qquad (9)
$$

Expressions for the coefficients a_{ij} and b_i $(i,j = 1,4)$ are given in Table 1. In state space system notation equation (9) is :

$$dT/dt = UT + VTG \qquad (10)$$

where $U = (a_{ij})$ is a 4×4 matrix and

$$T = [T(A),T(B),T(C),T(D)]' \text{ and } V = [b_1,b_2,b_3,b_4]'$$

are 4-vectors.

Equations (9) are ideally suited to the on-line determination of the cold spot temperature $T(A)$ since they require only the off-line calculation of 16 coefficients $(a_{14} = a_{23} = a_{32} = a_{41} = 0)$ and iterative solution from known initial conditions using the measured input temperature TG . Additional internal coil temperatures $T(B)$, $T(C)$ and $T(D)$ are also evaluated in this method.

4. MODEL ACCURACY

The above method for determining the cold spot temperatures in coils during annealing cannot be validated by the traditional techniques of numerical analysis since its accuracy depends largely upon the accuracy of the quartic temperature prfile assumption. Consequently the method has been tested by comparing cold spot temperature-time histories as simulated by the detailed simulation model with the temperature-time histories predicted by equations (9) using the gas stream temperature TG from the detailed simulation. The comparison has been made for each coil in a number of charges.

The initial test runs showed that while the predicted cold spot temperature-time histories had the correct form, the estimated temperatures were too high.   This error was easily minimized by varying the assumed fractional cold spot radii (which are the free parameters of the model) within known acceptable limits and after four test runs the results plotted in Figures 4 to 6 were obtained.   Each one of Figures 4 to 6 shows the cold spot temperature prediction errors (predicted temperature minus temperature calculated in detailed simulation) for all the coils in a single charge.   With one exception (coil 3 in Figure 4) the prediction errors of all the coils considered are very similar, being less than 10°C during the critical periods towards the ends of the heating and cooling sections of the annealing cycles.   In fact most errors are less than 10°C for the duration of the heating periods.   The sudden increases in all errors at the beginning of cooling cycles are caused by a decrease in the radial conductivity due to coild bore wraps contracting away from their neighbours (the gradient $\partial T/\partial r$ is reversed) which has not been accounted for in our simple model but could be by evaluating and using different coefficients a_{ij} and b_j , i = 1,...4 , j = 1,...4 , during cooling.   It should also be noted that the larger error in the prediction of the temperature during heating of coil 3 in Figure 4 is of little practical significance since the coil was hotter than the others and its temperature would therefore not affect cycle timing.

The above tests demonstrate the great accuracy of the simple model.   The ease with which parameters were adjusted to achieve this confirms the belief that the most useful first model of a complex system is that with the least number of parameters, based upon the greatest number of reasonable simplifying assumptions.

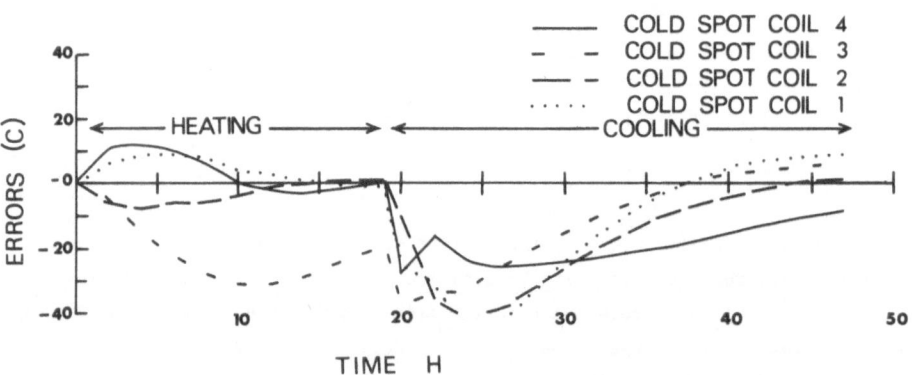

FIGURE 4   PREDICTION ERRORS

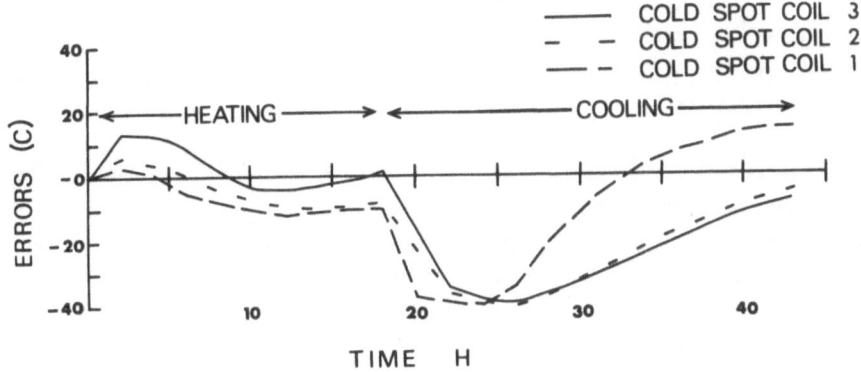

FIGURE 5 PREDICTION ERRORS

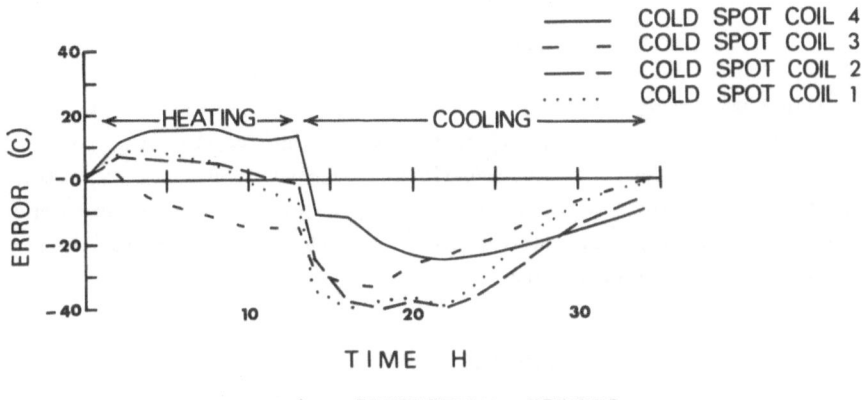

FIGURE 6   PREDICTION ERRORS

5. FURTHER MODEL REDUCTION

The performance of the quartic approximation based model naturally leads to the question as to whether an even simpler model with similar prediction accuracy can be obtained. In terms of on-line computation speed and storage nothing will offer much advantage over the quartic model but a model with fewer coefficients characterizing each coil in a charge could be used in a charge building algorithm which selects coils with similar heating rates for incorporation into the same charge, thus minimizing time wastage and product quality variation.

The simplest possible approximation to the heat equation arises when the quartic temperature profile assumption is replaced by the assumption that profiles are quadratic i.e. in equation (3) c = f = 0 .   In place of equations (9) this gives :

$$dT(A)/dt = \alpha(TG-T(A)) \qquad (11)$$

A single coefficient, α , characterizes a coil.   Unfortunately, despite all possible parameter variations, the predictions of (11) cannot be made to match detailed simulation results with sufficient accuracy - errors of over 60°C always occur during early stages of heating and frequently these never reduce to below 20°C.

The next logical choice of approximation, by a quadratic profile in one direction (axial) and by a quartic profile in the other direction (radial), has not been evaluated. Instead, the possibility of improving the predictions of equation (11) by introducing a time delay Δ has been investigated, i.e. (11) is rewritten as :

$$dT(A)/dt = (TG-T(A))\big|_{t-\Delta} \tag{12}$$

Accurate predictions similar to those given by the quartic model have been obtained in this way and methods for estimating suitable time delays Δ are currently under study (see [1]).   The two coefficient model (12) appears suitable for use in a charge building algorithm.

6. CONCLUSION

A very simple model suitable for on-line internal coil temperature estimation has been developed using theoretical and experimental furnace and charge temperature behaviour to reduce the complexity of the heat transfer equations.   A powerful determining factor in the choice of approximation was the limited availability of temperature measurements under normal production conditions.   Attempts at model reduction show that no further simplifications can be introduced without compromising accuracy.

APPENDIX 1

PROPERTIES OF SYMMETRIC QUARTIC FUNCTIONS

Consider a quartic function f of x , symmetric about $x = 0$.

$$f(x) = a + bx^2 + cx^4 \tag{A1}$$

The coefficients a , b and c can be expressed in terms of the values of f at points $x = n\Delta$, $n = 0,1,2$.

$$a = f(0)$$

$$b = 1/12\Delta^2[-15f(0) + 16f(\Delta) - f(2\Delta)] \tag{A2}$$

$$c = 1/12\Delta^4[3f(0) - 4f(\Delta) + f(2\Delta)]$$

Similarly if $h(x,y) = f(x)g(y)$ with f a symmetric quartic function, $f(x) = a + bx^2 + cx^4$ then for any y and Δ

$$a = h(0,y)/g(y)$$

$$b = 1/12\Delta^2 g(y)[-15h(0,y) + 16h(\Delta,y) - h(2\Delta,y)] \tag{A3}$$

$$c = 1/12\Delta^4 g(y)[3h(0,y) - 4h(\Delta,y) + h(2\Delta,y)]$$

Derivatives of h with respect to x can be evaluated as follows :

$$\partial h/\partial x = [\partial f/\partial x]g(y)$$

$$= [2bx + 4cx^3]g(y)$$

$$= 1/6\Delta^4 [\Delta^2 x \{-15h(0,y) + 16h(\Delta,y) - h(2\Delta,y)\}$$

$$+ x^3 \{6h(0,y) - 8h(\Delta,y) + 2h(2\Delta,y)\}] \tag{A4}$$

$$\partial^2 h/\partial x^2 = \partial^2 f/\partial x^2 g(y)$$

$$= [2b + 12cx^2]g(y)$$

$$= 1/6\Delta^4 [\Delta^2 \{-15h(0,y) + 16h(\Delta,y) - h(2\Delta,y)\} \tag{A5}$$

$$+ x^2 \{18h(0,y) - 24h(\Delta,y) + 6h(2\Delta,y)\}]$$

At points $x = n\Delta$, n = 0,1,2, these expressions simplify considerably, for example :

$$\partial h/\partial x\big|_{x=2\Delta} = 1/2\Delta[11h(0,y) - 16h(\Delta,y) + 5h(2\Delta,y)] \tag{A6}$$

$$\partial^2 h/\partial x^2\big|_{x=\Delta} = 1/6\Delta^2[3h(0,y) - 8h(\Delta,y) + 5h(2\Delta,y)] \tag{A7}$$

Equality in (A6) and (A7) is strictly true only when h is quartic in x but the right hand sides may be considered as approximate expressions for the derivatives of any function h for which $\partial^2 h/\partial x^2 > 0$ for all x or $\partial^2 h/\partial x^2 < 0$ for all x .

APPENDIX 2

COIL BOUNDARY CONDITION

Coil boundary heat transfer conditions have the general form :

$$k_n \frac{\partial T}{\partial n} = \sum_i q_{ni} \tag{A8}$$

The right hand side of (A8) is the sum of all radiative and convective heat flows into or out of a unit area of coil surface.

All heat transfer to the coil bore is by convection from the circulating annealing gas (which has negligible emissivity).   Hence in the bore :

$$- k_r \partial T/\partial r = h_r (TG-T) \tag{A9}$$

Since the variations of the conductivity k_r and the convection coefficient h_r with temperature are similar both coefficients can be regarded as constants.

At the coil end the total heat flow can be expressed as

$$q = k_1 (TC-T) + k_2 (TC^4-T^4) + k_3 (TG-T) \tag{A10}$$

The convector plate has a small thermal mass and can be regarded as being in thermal equilibrium with its surroundings, so that its temperature TC is specified by :

$$k_1 (TC-T) + k_2 (TC^4-T^4) + k_4 (TC-TG) = 0 \tag{A11}$$

By linearizing the radiation terms in (A10) and (A11) about a nominal temperature $TN(=450°C$ for example) at the coil end (A10) is :

$$- k_z \partial T/\partial z = q = h_e (TG-T) \tag{A12}$$

where
$$h_e = k_3 + k_4 k_5/(k_4+k_5)$$

and
$$k_5 = k_1 + 4TN^3 k_2 .$$

REFERENCES

[1]   Gruca, A. and Bertrand, P. Approximation of High Order Systems by Low-Order Models with Delays. Int. J. Control, 28, 6, 953-965, June 1978.

[2]   Harvey, G.F. Mathematical Simulation of Tight Coil Annealing. J. Aust. Inst. Metals, 22, 1, 28-37, March 1977.

[3]   Roberts, W.L. Cold Rolling of Steel. Marcel Dekker, 1978.

[4]   Thelning, K.E. Steel and Its Heat Treatment. Butterworths, London, 1975.

[5]   Van Asperen, G. Metallurgical Considerations when Annealing Large Coils. Mechanical Working & Steel Processing Vol. 11, pp.13-32, AIME, 1973.

INDEX